BYE BYE SATAN

Living by Evidence

&

Bruce Oakley

Professor Emeritus
Molecular, Cellular, and
Developmental Biology

"We must find God in what we know, not in what we do not know."
Dietrich Bonhoeffer 1906-1945

This book is dedicated to the memory of Dietrich Bonhoeffer, a Lutheran theologian of undaunted courage. Even today, Bonhoeffer's thoughtful insights resonate in theological circles (Bethge, 1972). At great personal risk, he spoke out and sought to bring down the Nazi regime. Scarcely one month before victory was declared in Europe, Dietrich Bonhoeffer was stripped naked in his cell, marched outside, and hanged by the SS in the Flossenbürg concentration camp on the morning of April 9, 1945.

Order this book online at www.trafford.com/07-3128
or email orders@trafford.com

Most Trafford titles are also available at major online book retailers.

Note for Librarians: A cataloguing record for this book is available from Library and Archives Canada at www.collectionscanada.ca/amicus/index-e.html

Printed in Victoria, BC, Canada.

ISBN: 978-1-4251-6772-1

We at Trafford believe that it is the responsibility of us all, as both individuals and corporations, to make choices that are environmentally and socially sound. You, in turn, are supporting this responsible conduct each time you purchase a Trafford book, or make use of our publishing services. To find out how you are helping, please visit www.trafford.com/responsiblepublishing.html

Our mission is to efficiently provide the world's finest, most comprehensive book publishing service, enabling every author to experience success. To find out how to publish your book, your way, and have it available worldwide, visit us online at www.trafford.com/10510

www.trafford.com

North America & international
toll-free: 1 888 232 4444 (USA & Canada)
phone: 250 383 6864 • fax: 250 383 6804 • email: info@trafford.com

The United Kingdom & Europe
phone: +44 (0)1865 487 395 • local rate: 0845 230 9601
facsimile: +44 (0)1865 481 507 • email: info.uk@trafford.com

10 9 8 7 6 5 4 3

Contents

Preface

Bye Bye Satan arose out of the tumult of university seminars. I was touched by the spirituality of students and their concern over the clash between evidence and faith. In mirroring their family backgrounds, the students reflected the remarkable diversity of religious views in America. Our free-flowing discussions generated some common understandings found in this book. Still, the search for truth and consensus turned out to be rather more than a semester-long endeavor. Heartlessly, I informed them it might be a life-long project to square one's beliefs with the abundance of new knowledge coming along.

As you can sense from this book's title, I would very much like to see humans unburdened from the gloom of mean-spirited contrivances such as Hell. Adults ought to aim higher than grim fairytales. *Bye Bye Satan* points out some past damage, and present and future threats, from governments and institutions propelled to act on faith rather than on fact. My recounting of some of the dust-ups between evidence and faith may suggest a winner-take-all tussle between science and religion. However, it would be throwing the baby out with the bathwater to interpret my outright rejection of the abuses of faith as the rejection of all religion. Both science and religion have their strengths and their shortcomings. My intention is to present a candid account. Religion is the more frequent lightning rod for criticism because its traditional or fundamentalist stances include intolerant rigidity that resolutely distorts ancient history while denying central scientific advances. It is not anti-religious to reject fanciful scriptural accounts and to condemn callous treatment by clerics determined to elevate their religious institutions at the expense of human lives on Earth. Science also comes in for criticism, since every human endeavor has episodes of human weakness or good intentions gone awry.

Religion is concerned with the great questions of meaning—with why. Science is mainly concerned with questions of mechanism—with how. Religion is free to pose many extraordinarily fundamental questions about ultimate meaning and the great beyond that science can't address. Although this limitation of science clears a path for spiritual questing, it regrettably leaves theologians without the most effective proven methods for rooting out truth.

Three broad American audiences of *Bye Bye Satan* are likely to have different reactions. The first group comes from a tradition that considers scripture to be divinely inspired, perhaps even inerrant, but more likely accurate unless credible research has shown otherwise. Those in the second group were raised in a more modern theological tradition that considers scripture inspiring, but not the inspired Word of God. While believing in a God, they have their doubts about the reality of other supernatural beings and destinations like Limbo, Purgatory, Hell, and perhaps Heaven. Those in the third group are not members of a religious denomination as ordinarily defined by communal worship and doctrinal belief. They have a personal secular or spiritual outlook arrived at either by informed choice or because they have never grappled deeply with religious issues. Where you stand at the finish may depend upon where your loyalties stood at the start. Most of us remain attached to the religion of our childhoods.

Surveys indicate that 60% of Americans consider religion to be very important in their lives. Over 90% believe in a higher power, 67%-75% pray daily, 40% attend a church or synagogue regularly, and 82% seek spiritual growth (Lee and Newberg, 2005). A Scripps Howard News Service poll in the United States found that although 72% of adults believed in an afterlife, only half thought they would be physically resurrected. It would be interesting to know where they believed their resurrected bodies would go and how they would survive. Perhaps, because of diverse doctrinal beliefs in pluralistic America, some consider it impolite to question a person about their religious beliefs. Silence is the preferred mode of religious tolerance. My students found this somewhat odd, since at the least you have a moral duty to question those whose activism impinges unacceptably upon lives you value. Surely, one should question at length the faith-based advocates of violence and torture. If one absolutely believes in a particular religious doctrine because one has gotten faith, then all differing opinions necessarily must be wrong. In tolerating other creeds the faithful are more likely to exhibit pained or pious silence than openness to diverse views or searching analyses like the one about to unfold.

Some adults experience epiphanies that move them from fundamentalist preacher to anti-religious zealot (Barker, 1992). Others come on board for the reverse trip from skepticism to evangelism (Collins, 2006). By nimble compartmentalization they may embrace both religious faith and scientific truth. Even so, the same flood of feelings that carries one to faith can drown valid evidence when the two compartments inevitably mingle. Where you stand on the continuum of religious belief will reflect the clarity of your discernment arising from your particular religious background and your commitment to thoughtful contemplation.

Life has many exasperating intrusions that steal away time we might use for contemplation. With advertising and email spam leading the onslaught, we are subjected to a morass of manipulative drivel. Don't we warrant some quiet time to sift through the facts about life on Earth for ourselves? In taking stock we need to beware of skewed and superficial advice from cryptic sources on the Internet, or "cherry-picking experts" who find it marvelously convenient to avoid mentioning facts that don't fit. Happily, it is often easier to "get up to speed" than you might think. If you arouse yourself to explore a subject on your own, you can develop enough familiarity to double-check the experts. Admittedly, even when we have the expertise, we rarely take the time to check out the strength of the evidence behind "truths" we are told, especially if it is what we wanted to hear. My primary purpose is to make your assessment of life on Earth easier by bringing together in one book a frank discussion of religion in the context of modern science, or if you prefer, science in the context of ancient and modern religious practices. A book that compares science and religion, with special emphasis on their intersecting concerns, risks receiving the same reaction as a judge's ruling in a contested divorce—at the close both parties feel dissatisfied with the distribution of assets. That is why it is important for you to read and probe in order to make up your own mind through critical analysis, rather than passively accepting the dictates of any authority, myself included. If I could inspire you to a clearer view than mine, it would be a sublime reward. No educator could ask for more.

I write with the conviction that the spiritual wings of all humanity are lifted by the same basic concerns we will consider here. If so, there is hope that universal values will emerge and crystallize out of the ongoing accumulation of facts. A shared collection of facts and premises may lead to co-operative solutions for planet Earth. We need them. But we can only find common ground when we are well-informed and willing to listen to each other. Closed ears and closed minds make continuing conflict inevitable. Even as understanding and enlightenment accrue, it is hardly surprising that some individuals will cling to ancient views and superstitions, even claiming a holy right to such rigidity. But preferably we will engage in continual re-examination, and react positively when new findings open new vistas that clarify understanding. We ought to resist the tendency for settled feelings about our assessment of life on Earth. As they always have, differences in human disposition and rearing will fuel debates about the legitimate or optimal path in life. In an era of terrible weapons and intensive environmental exploitation it is critical for bona fide understanding and compassionate hearts to prevail. We must contemplate our situation carefully if we are to survive. If the powerful are persistently ignorant, they will lead humans into catastrophic conflicts.

Experienced readers may feel that I needlessly rehash well-known unpleasantness like the Inquisition, or that I set out straw men only a naïve person would entertain, like believing that Heaven is something you can almost touch right up there in the clouds. Mainstream Christianity has come so far in the last few centuries that it may seem unnecessary to recount earlier harsh actions or bogus factual claims. One purpose in recounting such historical failings of Christianity is to immunize those less well-informed. Moreover, are we really at no risk of a new dark age of authoritarianism? In the view of many observers, pompous politicians are currently mocking America's traditions of liberty and freedom by shredding the Constitution's checks and balances and defiling America's vaunted commitment to human rights and due process under the law. Lastly, my recounting of biblical history and classical arguments intends to place a diverse readership on the same footing. Because it is too often brushed aside, it seems particularly important to recount the difficulties that the existence of human evil and suffering creates for the God of the New Testament.

You are correct to sense that some of my core moral beliefs will leap forth into the open. I can tell you that these concerns focus on the importance of an informed reverence for life, and deep and wide apprehensions about the arrogance of authoritarianism, hypocrisy, abuse of children, and torture. I expect there will be general agreement on these issues but marked differences in the intensity of concern. My uneasiness with the arrogance of power helps to explain why I am at odds with Fundamentalist and rigid evangelical sects. Many of us would rather have awkward truths prevail than agree to an ancient code riddled with fictions.

The primary aim of this book is to urge its readers to lead an evidence-based life. If you have read a variety of books in the broad subject area of life's origins, evolution, human nature, religion, evil, and the like, you will have encountered many excellent works. Nevertheless, they do run the gamut. An essay's tone may be sad, exhilarating, thoughtful, evasive and manipulative, superficial, intellectually draining, insightful, haughtily dismissive of religion or science, mere pandering commercialism, even intentionally demeaning, or worst of all, persistently unclear, like some gaseous writings on legal philosophy. I have specifically attempted to avoid using debating techniques that demean the reader as in, "If you don't believe Jesus was sincere in saying this, then you must believe that he was a lunatic." (Gasp.) However, in several instances I let it slip that a proposal seems truly dim, or I show deep concern about particular treacherous paths in life, especially those that point toward absolutism.

In a book that marches over the terrain of beliefs truly important to humans—often bloodied by fiercely contested wars—readers are entitled at the outset to a better sense of the author's perspective. With a Ph.D. focused on neuroscience followed by 40 years as a biologist investigating the developing mammalian nervous system, I am comfortable with issues of nature and nurture and life on Earth. As you might expect from my background in neuroscience, I believe that no issues of human feeling and spirituality make sense except in light of brain function. Since I comment frequently about religion, you should understand my religious background. I was raised in the Quaker tradition that relies primarily on each person's Inner Light for guidance rather than upon creeds and scripture and a hierarchical church. Perhaps an individual uncommitted to doctrinaire beliefs is freer to make even-handed appraisals of religion, letting the facts lead the way. I have tried to do this.

For multiple reasons you may wish to pursue particular topics further. In matters of science and church history the facts are generally available in the articles and books I have listed, or in standard references. Details are sparse for the events that occurred before and during biblical times. By definition one departs from the trail of facts in moving beyond knowledge to faith. What is one to do when there is sparse documentation? Some of many undocumentable claims, like what spurred religiousness in prehistoric times, reflect the prevailing views of anthropologists and sociologists. Frequently, I will leave a contentious issue as an unsolved puzzle rather than force a conclusion. Patience is a virtue. It may take scholars decades of research to resolve historical disputes. Ultimately, I hope to provide the satisfaction and security that comes from knowing that together we honestly assessed some perennially challenging issues with new evidence and fresh comparisons. By helping to organize your further inquiries, this book may inspire you to join in constructive responses to our situation on Earth. Persevere in making the world a better place and be guided by the advice of the quietly courageous Quaker, A. J. Muste, and "Speak Truth to Power."

I am indebted in multiple ways to the following individuals for their advice and commentary: Bruce Carlson, Johnny Palka, Phil Clampitt, Carol Blotter, Sarah Frommer, Fred Mayer, David Mindell, Rev. Terry N. Smith, Andy Balash, Ehren Brenner, Brian Boguslawski, Grace Butts, Mike Coulter, Hayden Gandolfi, Ashwin Gupta, Cory Hafer, Lisa Haidostian, Carly Hanson, Matt Koletsky, Lisa Kurajian, Stephanie Levy, Prash Mahalingam, Amy Mason, Anneli Purchase, Carol Rankin, Aaron Sandusky, Tyrone Schiff, Delsia Vernon, and Josh Appleman.

1

&

Religion and Science: Quests for All Seasons

Religion Occupies Many Lives

The search for inspiration in religious writings is a spectacularly popular quest. With a remarkable showing of more than a million titles in print, books on religion aim to provide comfort and to inspire belief. Thousands of authors touting the Christian religion offer a wide assortment of doctrinal visions. Numerous religious books and associated videos are produced by the multi-million dollar budget of a subculture of Fundamentalist Christians which aims to immunize their youth against the threats of science. A tiny countering genre of books dismisses religion. I try to avoid these extremes. This is a book about the tenets and methods of science and religion, and the clash between evidence and faith. My responsibility is to offer up a main course of information seasoned with questions that contribute to your spiritual growth. We will explore both the spiritual and biological facets of life on Earth. The discussion of religion is set in the context of Christianity. The context of science is the DNA code and the evolutionary basis for life's diversity. It is a sumptuous feast that I try to serve in palatable portions.

The Bible is the preeminent book for Christians—it is their spiritual guide. It is ancient but not chronological. The Old Testament leads off with Genesis, although those sections were actually compiled centuries after the narratives of Amos, Isaiah, and

Jeremiah. Because Genesis was written several thousand years ago, it is scarcely surprising that it misses the mark on some matters. After all, no one claims the Bible is, or ever was, a scientific textbook. In no sense was it a systematic compendium of up-to-date technical information. We read that the earth was initially covered with water before any dry land appeared (*Genesis 1:9*) and the earth formed before the stars (*Genesis 1:16*). For each claim the sequence needs to be reversed. Formed after the sun, the surface of planet Earth was initially hot enough to convert all liquid water into steam.

Even though many denominations claim to take the Bible literally, in practice they cherry-pick the most palatable sections and employ elastic interpretations stretched to fit their needs. Strict fundamentalist groups are Creationists in the sense that they believe the claim in Genesis that God created the universe, the earth, and all of its life in six 24-hour days some 6,000 years ago. Jehovah's Witnesses disagree. They believe God created these things over a time span compatible with the findings of modern cosmology, geology, and biology. Although properly rejecting the erroneous geology and cosmology of Genesis, Jehovah's Witnesses deny the gradual descent of organisms, one from another. For them "evolutionary theory and the teachings of Christ are incompatible" (Awake magazine, September, 2006, p. 10). Most tellingly they deny that man evolved from apes since "God created man in his own image..." (And surely God was not an ape!) Nevertheless, many Catholic, evangelical, and Protestant biologists believe not only in the scientific account of the creation of the universe and Earth, but also in the evolution of its life according to principles that Charles Darwin first formulated (Alaya, 1998; Miller, 1999; Falk, 2004; and Collins, 2006 among others).

With many young people caught in the swirl of literal interpretations of Genesis and the supposedly horrific implications of evolution, it is important to outline the relevant evidence beginning with Chapter 2. One must be willing to hold up venerable religious beliefs to the light of new understanding. Many centuries ago a wise St Augustine (354-430 CE) wrote about his concern over shortsighted, defensive strategies, "It is a disgraceful and dangerous thing for an infidel [non-Christian] to hear a Christian, while presumably giving the meaning of Holy Scripture, talking nonsense. We should take all means to prevent such an embarrassing situation, in which people show up [the] vast ignorance in a Christian and laugh it to scorn...how [then] are they going to believe those Scriptures in [the more important] matters concerning the resurrection of the dead, the hope of eternal life, and the kingdom of heaven?" (cited in Ayala, 1998).

When I explore traditional religious practices and doctrines as responses to valid spiritual needs, it will be in the context of evolution. There is nothing in biology that lies outside the scope of evolution. Questions about human evolution like "Who are we?" and "Where did we come from?" are addressed not only by ancient religious narratives but also by new genetic information, like the astounding similarity between the millions of letters of genetic code in human and chimpanzee DNA (deoxyribonucleic acid). In focusing on the evolution of life on Earth and the roots of religion, *Bye Bye Satan* draws from modern science, including molecular genetics and neuroscience. The unfolding revolution in molecular genetics offers multiple benefits and challenges to health and reproduction. And neuroscience, the study of brains, attracts attention at levels ranging from the dazzling properties of nerve cells to the relentless withering of diseased brains.

Especially informative is the molecular genetic evidence for the relatedness of living organisms as one traces back to the common ancestors shared in the distant past. From cosmology to chromosomes, modern science impacts religion in diverse ways. Having experienced first hand the revolution in molecular biology, I am frequently awed by the scope of deepening clarifications about the mysteries of life. The ongoing deluge of important new findings directs us to make informed evaluations of spiritual as well as material beliefs. In this account I summarize some religious implications of the news from cosmology, chemistry, paleontology, and biology—and most especially from molecular genetics. It is of considerable interest to see which religious leaders will respond positively to the fresh evidence that clarifies and invigorates the life sciences.

I use the tenets of Christianity and the construction of the Bible to illustrate how human needs have driven religious doctrines. I emphasize the Bible because it is a familiar spiritual gateway. No less interesting are the revelations of several recently examined Gospels that were excluded from the New Testament. In addition to religion's contribution to basic human spiritual needs, I note how the institutional operation of major religions shapes religious practice. In trying to tell the truth about life on Earth, this book may conflict with your cherished vision of Christian practice. Of course, scattered differences in viewpoint and emphasis are inevitable, but if there are strong and broad conflicts with the substance of scientific conclusions, it would be wise to sort them out.

Generating evidence-based information is the hallmark of science. Scientists make discoveries when they are alert to surprises and when their informed curiosity makes them eager to explore. Peter Medawar won the Nobel Prize (1960) for studies on the immunology of transplantation and the body's acceptance of grafts of foreign tissues. He believed that philosophers of science who used physics as a model have had too narrow a view. There is more to the scientific method than the testing of hypotheses. Without testing any formal hypothesis, careful observations and descriptions of the universe and life on Earth have made enormous contributions to understanding. Prior to optical instruments the unaided eye was the only way to make visual observations. With the invention of telescopes, microscopes, spectroscopes, and other light-gathering devices the remote and the tiny could now be seen—new understanding flowed from improved observations.

Later on we consider what it means to test a hypothesis and draw conclusions. For now let us stress the crucial importance of being able to repeat observations with the same result—the process of replicating the findings. Hence, an important strength of the scientific method lies in allowing other scientists to check out your findings. To do this, others must be given a clear and orderly sequence of methods to be followed. Magicians, conjurers, and other charlatans have always been keen to conceal their methods. By contrast scientific findings must be open to independent verification, or replication by others who follow the original methods.

Modern scientific advances depend mainly on the creative follow-up of recent findings with the use of powerful new methods. As every experienced scientist has come to appreciate, if you restrict yourself to old methods, it is unlikely you will make observations more astute than those of scientists of earlier eras. Scientists today are not inherently smarter than investigators of previous generations. Of course, present day scientists rely upon a far more extensive base of accumulated understanding. Just look at the increased thickness of current textbooks. But the modern scientist's greatest advantage is a wide range of stunningly powerful new methods.

New methods and instruments allow scientists to investigate life on Earth and discover facts that engineers may incorporate into various applications. Most technological advances reflect incremental improvements that make an existing invention still better—like finding a better inert gas to fill a light bulb. Typically, after some initial insight, the development of any new technology such as light bulbs, phones, radios, air brakes, and diesel engines, reflects the efforts of many individuals whose hard work yields successive improvements that ultimately convert a clever idea into a practical application. Some advances, like the discovery of nylon and penicillin, were wholly unexpected outcomes whose potential value was recognized by an astute observer. Creative, well-informed minds may blaze their own paths to discovery. It deserves mention that not all commercialized applications are true advances—it is commonplace for a standardized product to be tweaked or repackaged merely to sell more of essentially the same device or drug.

Scientists aim to assess nature with detached objectivity. But it is a challenge to remain neutral if one is intensely invested in the process. Zest, egoism, ambition, and envy can get in the way. There will always be idiosyncratic or outlier scientists who steadfastly resist accepting secure conclusions. On the other hand, popularity is not a guarantee of scientific truth, for occasionally it is the majority who draw flawed conclusions from thin evidence. Once widely held but erroneous views have included a non-existent gas called phlogiston, phrenology (reading one's personality from bumps on the skull), vitalism (an undetected spirit whose entry brings cells to life), aspects of Freudian psychoanalysis, and personal stress as the cause of stomach ulcers (see Chapter 11). It is human nature to believe what we are told, in spite of scant evidence in favor or even telling evidence to the contrary. Both because religion is a more difficult endeavor and because it settles in so closely against our hearts and feelings, reasoned critiques of religion are all the more anguishing. Of course, in the long run it is a risky strategy to hide from factual analyses. Remember: "Fiction in—Friction out."

Religion has the challenging objective of comforting our hearts with wise insights. Religions must address three gnawing issues of existence: the birth of the universe leading to a diverse spectrum of life on Earth, the presence of evil and suffering, and the inevitable death of individual living creatures. Birth, torment, and death are the existential realities that guarantee religious contemplation. Under its own billowing tent, science shelters cosmologists and astronomers who examine the origin of the universe with its massive objects and prodigious energies, as well as chemists and biologists who probe the origins and properties of life on Earth. Many others, including the health and legal professions, governments, and charities, cope with suffering and evil—the many torments and injustices in life. Perhaps, before the reader and I grapple with these weighty subjects, we should ground ourselves by looking at the origins of the cosmos and life.

2

The Origins of the Universe and Life on Earth

Grasping at the Cosmos and Life

Gaining an intuitive understanding of the scale of our universe is like trying to take in the night sky by looking through a soda straw. Not only is the universe spectacularly broad and deep, it is also astoundingly ancient. Consider the history of your own chemical ancestry. Your molecules go far back to an expanding universe that created trillions of galaxies (each a dispersed cluster of millions of stars) including our particular galaxy of stars that continue to make matter and spew it out. One gaseous nebula condensed into our own sun, along with fragments that later congealed into Earth. The bodies of today's living animals and plants are built from atoms that were formed billions of years ago by solar nuclear fusion. We are each composed of chemical atoms far older than the first life on planet Earth.

Although some cosmologists imagine a stupendous number of universes in addition to the one we live in, we will focus on our own universe with its many unresolved scientific questions. Theoretical physicists believe there is a complex harmony in the distant and dimly understood features of our universe. Ninety-five percent of its mass is cold, dark matter that surrounds the galaxies. Galactic light is distorted as it passes through the gravitational field of dark matter. When the distortions of galactic light are spatially mapped it reveals that dark matter occurs in stupendous cable-like networks. In the primeval universe dark matter probably acted as a gravitational scaffold that attracted stardust into the aggregates we see as today's galaxies. Near the center of most galaxies is a super-massive black hole that contain heavens-knows-what since its awesome gravitational pull engulfs all matter that comes too close to it even as it spews out energy.

Earth is but a tiny speck in a colossal expanding universe that began 14 billion years ago with an unimaginably rapid initial expansion called the Big Bang. (This was a cleverly mocking term coined by the eminent British astronomer, Fred Hoyle. Ironically, Hoyle was skewered by his own sharp wit when others actually found the phrase appealing and adopted "Big Bang" as a popular shorthand.) The Big Bang started as an expansion of a tiny point where all space, time, and mass may have been collected as if compressed. The recipients of the 2006 Nobel Prize in Physics, John Mather and George Smoot, led a team that used NASA's COBE satellite (COsmic Background Explorer) to measure ancient microwave radiation just now reaching us from the most remote regions of the universe. In support of the Big Bang hypothesis, they detected lumpiness and non-uniformities just as predicted from the concept of galaxy formation.

An unsatisfying answer to the natural follow-up question "What surrounded the spot where the Big Bang began?" is that there was nothing. So if the Big Bang is hard to envision, what might have preceded it is even harder to fathom. And there must have been a trigger for the "explosion." One would think the blob of energy/matter surely came into existence from something. In searching for ultimate causes one could imagine that the energy or the material of the universe may have existed forever, perhaps endlessly cycling between expansion and contraction, or that it emerged from the Big Bang as an uncaused event, or even arose from nothing. Existing forever? Cycles without a beginning? Uncaused events? Something from nothing? These notions are profoundly baffling. Similarly difficult questions confront the supernatural creation myths of religions. What was God doing before he created the universe and when and how did God arise? It seems to this non-cosmologist that scientific and supernatural approaches are both stymied in trying to explain what went on before the universe began. However, an unstructured pre-universe with either dispersed or highly compacted energy and matter is less perplexing than trying to figure out what sophisticated process could have constructed a God who was exceeding complex before the birth of the universe.

Rather than struggling with an endless regression toward a first cause or puzzling about the complexities of the initial expansion, it is easiest if we begin with the state of the universe about 380,000 years after the Big Bang. By that time the universe had "cooled" enough to permit the abundant protons to "absorb" and thin out the dense fog of free electrons. As protons and electrons combined into neutrons, it cleared out paths for particles of light (photons) to travel without being dimmed by a dense fog of free electrons. Increasingly unimpeded, these photons were able to travel astounding distances.

The closest star to us, apart from our own sun, is Proxima Centauri. It is a mere four and half light-years away. This means if you traveled at the maximum possible speed—the speed of light (186,000 miles a second) you would get there in four and a half years. Traveling at the typical speed of a spaceship (7 miles a second) you would get there in about 100,000 years. Should Proxima Centauri emit a burst of photons while you read this, it would take four and a half years for those photons to reach Earth and allow us to see the flash. Similarly, light arriving today from a remote galaxy 10 billion light-years away, must be composed of photons 10 billion years old! Imagine looking toward a distant galaxy in the night sky and suddenly seeing an exploding star, the telltale flash of a supernova. It is eerie to realize you are actually seeing events as they were happening some millions of years ago. Whoever said we are trapped in the present when our eyes can feast on such ancient events?

The Deep Field experiment in 1995 had the Hubble telescope point toward one site in the sky for 10 days in order to gather the faint light of thousands of galaxies—a deep sample of the more than 200 billion galaxies in the universe. A galaxy might be 200,000 light-years across and hurtling through space at a speed of 555 miles a second. Our own "milky way" galaxy contains 100 billion stars including our own sun positioned about two-thirds of the way out one arm of our spiral galaxy. To supply some context, 100 billion stars is equal to the number of grains of wedding rice it would take to fill a large cathedral like St Peter's basilica in Rome.

The scale of cosmic space and time is so immense that even professional astronomers struggle to connect intuitively with the vast number of distant planets dispersed in the universe. Try this on your next visit to a beach. Pick up a single grain of sand. Pretend that one grain is planet Earth. Then try to imagine all the grains of sand that make up this beach as well as all of the other ocean beaches across our entire world. This is about the number of planets believed to be orbiting the many trillions of stars in the known universe. Many of these planets are frozen solid; others are so hot they are molten. Then again, if planets are as common as grains of sand on all of Earth's beaches, it is likely that some planets have conditions suitable for the evolution of life as we know it. Life is most likely to arise if a planet has mild temperatures, an atmosphere, liquid water, and several crucial elements such as carbon and nitrogen. You should thank your lucky stars for the privilege of living on one such accommodating planet, especially at this moment—an era of unparalleled understanding.

> To see a world in a grain of sand,
> And a heaven in a wild flower.
>
> (William Blake, 1757-1827)

From the perspective of material comforts, these are extraordinary times. Only in the past century has humankind developed the technological base to ensure a comfortable life-style, at least for those in developed nations who can afford it. With sophisticated instruments the Human Genome Project and other genetic research have "read" several people's DNA in order to map the human genome—our genetic instructions. Are you aware what a fleeting moment it is to be alive at the instant the human genome is being decoded? The surging advances in molecular genetics offer a unique opportunity to be able at last to investigate the Code of Life. What a marvelously special time to be alive. Suppose we pretend to compress the 3.5 billion years life has existed on Earth into the time it takes to jog twice around an 18-hole golf course. On this intuitively manageable scale, the five-year-long international effort that decoded the human genome would be represented by the instant of time needed for the jogger's shoe to pass over a single blade of grass—our thin instant in time. These are times of awe and reverence for life, as we are poised on the threshold of learning how mammalian embryos develop and how the resulting specialized cells function. It is unfortunate that we lack comparable progress in quelling the fear and hostility in human nature. Similarly, our natural protectiveness about reproduction keeps us in deep denial that there are many more people than this crowded planet can sustain in peace and comfort. Even though there are still several billion fewer humans than the 10 billion projected to populate Earth, there is already a serious mismatch between the desires of people and resources—like the availability of inexpensive oil, pure drinking water, dietary nutrients, and medical care. If all such resources were suddenly distributed evenly among all people, everyone on Earth would be living much less comfortably than today's average American. You can sense the surge of denial when the instant response to this observation is to dream up new ways to accommodate yet more people. We should not confuse such noble attempts to accommodate more people with being on the path to sustainable solutions, let alone an optimal long-term solution.

How Special is Planet Earth in the Universe?

One of the important steps in gaining understanding is to be skeptical of common sense. "Just have faith in common sense" seems like harmless enough advice, except that it is frequently wrong. Consider the predicament of humans 5,000 years ago as they witnessed the daily apparent motion of the sun as it rose every morning in the east and seemed to move across the heavens to set in the western sky in the evening. It was further obvious to humans as they gazed upward that the stars also progressively moved westward in the night sky. It was reasonable to infer from such observations that the sun and the heavens above revolved around a stationary Earth. A further benefit of anchoring the earth at the center of the universe was that it nourished human pride. How human egos crave center stage. Both the apparent motions of the sun during the day and of the stars at night are illusions stemming from the earth spinning eastward on its axis. The spherical shape and true motions of the earth are non-intuitive. Both North Americans and Australians feel themselves as standing upright and stationary. A rapidly spinning Earth is especially non-intuitive since we don't experience a constant wind from the east—or more dramatically, feel ourselves moving, about to be flung off into space. The answer is that gravity and momentum keep both the atmosphere and us "stuck" to the same spot on this spinning globe. At roughly 1,000 mph, a supersonic plane or satellite headed east would remain directly above the same spot on the earth below. Even though as you read this you are traveling about 1,000 miles an hour, you don't feel the motion because our brains respond to changes in velocity, not to constant velocity.

Around 260 BCE the Greek astronomer Aristarchus (312-230 BCE) correctly proposed a heliocentric arrangement; that the earth and the other planets orbited around the sun. But this idea slipped into obscurity, overwhelmed by common sense and the ego-gratifying popularity of the geocentric view of Aristotle and

Ptolemy that the universe revolved around planet Earth. Understandably, mainstream Christianity adopted the same geocentric view some centuries later.

The celestial abode has had a complex organization in religious fantasies, not to mention the elaborate geometries in models by early astronomers. It was the brilliant St Thomas Aquinas (1225-1274 CE) who artfully arranged layers of fluttering angels into three hierarchies, each divided into three orders. Angels in the order of Powers moved the heavens, sun, moon, planets, and stars. The archangels protected religion on Earth while ordinary angels cared for other earthly affairs. Teams of wicked angels scooted about in swarms to generate lightning, storm, drought, and hail, all the while busily tempting men to sin. These bevies of good and bad angels stoked Christian life with supernatural intrigues nearly as boisterous as the passel of Greek and Roman gods during their frequent excursions into Athens and Rome. With some support from biblical texts, Aquinas' vast scheme espoused the Ptolemaic view that the sun revolved around a stationary Earth.

Subsequently, medieval Christian clerics continued to suppress public discussion of Aristarchus' long-abandoned heliocentric view until it was revived by the observations and calculations of Nicholas Copernicus in the 16th century, as set out in his book *Revolutions of the Heavenly Bodies*. Stymied by repeated delays in publication, Copernicus only briefly managed to hold and gaze at his completed book because by the time it was printed in 1543 CE, he was too near death to read it. Perhaps it was better that way. Unknown to Copernicus, the hastily revised preface of his book downgraded heliocentrism from a fact to a hypothesis because the publisher correctly feared church retribution. Even so, the book's message wasn't toned down enough, for the teaching of Copernican heliocentrism set the clerics to mutterings of "heresy." Ultimately, Anglican, Calvinist, Lutheran, and Roman Catholic Churches all vied for the leading role in denouncing heliocentrism as contrary to scripture, especially the 93rd Psalm that confidently yet vaguely asserts, "Yea, the world is established; it shall never be moved." Listen to Martin Luther as he dismissively commented on Copernicus, "People gave ear to an upstart astrologer who strove to show that the earth revolves, [rather than]...the sun..." Calvin also weighed in with his famed authoritarian flair, "Who will venture to place the authority of Copernicus above that of the Holy Spirit?" Later on, Catholic ecclesiastical authorities issued special orders that required oaths of secrecy from professors at important universities: Pisa in Italy, Innsbruck in Switzerland,

Louvain in Belgium, Salamanca in Spain, and others. Professors were sworn not to tell students what they were able to see with their telescopes.

The Institutional Church Confronts Galileo's Threat

Particularly concerning for the Vatican were Galileo's pioneering observations describing four moons circling Jupiter as their local hub (White, 1896). When he turned his attention to our own moon he saw valleys and mountains, which appeared to be illuminated by light from the sun, not by light created by the moon itself as scriptural interpretation had it. His observations of the shifting positions of sunspots suggested the sun was rotating. For all of this mischief Galileo was summoned before the Inquisition in Rome in 1615. The Inquisitors unanimously declared, "The first proposition, that the sun is the center and does not revolve about the earth, is foolish, absurd, false in theology, and heretical, because [it is] expressly contrary to Holy Scripture." Sensibly, Galileo beat a hasty retreat to Florence where he managed to continue his observations unobtrusively. However, fearing a vile fate, he published nothing. Even so, the forceful denunciations continued. In 1631 Father Melchior Inchofer, Society of Jesus, vigorously declared, "The opinion of the earth's motion is of all heresies the most abominable, the most pernicious, the most scandalous; the immovability of the earth is thrice sacred; argument against the immortality of the soul, the existence of God, and the incarnation, should be tolerated sooner than an argument to prove that the earth moves."

In 1632 Galileo published *Dialog Concerning the Two Chief World Systems*, which provided strong support for Copernican theory. Unable to react soon enough to suppress distribution of Galileo's book, an angry Pope Urban felt personally mocked. He immediately imprisoned Galileo and threatened him with torture unless he recanted. In his trial in 1633, Galileo was told his proof of heliocentrism was faulty. Some believe that Galileo remarked, "The Bible tells you how to go to Heaven, not how the Heavens go." This clever but strategically inept quip may have sealed his conviction on the puzzling charge of "vehement suspicion of heresy." The downside of this bizarre yet accurate charge was humiliating punishment. Galileo was forced to kneel down and make the following public confession:

"I, Galileo, being in my seventieth year, being a prisoner and on my knees, and before your Eminences, having before my eyes the Holy Gospel, which I touch with my hands, abjure, curse, and detest the error and the heresy of the movement of the earth." (Swearing on the Bible remains a ritual in American jurisprudence.)

Not yet done with burying this heresy, the church next distributed countering writings in favor of geocentricism while attempting to suck all mention of Copernican theory from textbooks and universities. This is the classic dual strategy of spin-doctors in any era: denigrate the truth and substitute fiction. For example, the cleric Scipio Chiaramonti wrote in regard to heliocentrism, "Animals which move have limbs and muscles; the earth has no limbs or muscle, therefore it does not move. It is angels who make Saturn, Jupiter, and the Sun to turn round. If the earth revolves, it must also have an angel in the center to set it in motion; but only devils live there. It would therefore be a devil who would impart motion to the earth…" Never underestimate the fanciful creations of human imaginations fueled by faith rather than fact.

Galileo lived for another 10 years under virtual house arrest while his prior observations and conclusions were systematically weeded out of all church-based colleges and universities in Europe. In sanitizing one scientific document that had characterized Galileo as "renowned" the Inquisition ordered the substitution of the word "notorious." Galileo died in 1642, the same year Isaac Newton was born. Three hundred and fifty years later in 1992 the Catholic Church got around to apologizing for condemning Galileo.

The validity of heliocentrism has gnawed at earthly pride and church formulations. In poetic affirmation of cosmologists, Alexander Pope accepted that the sun, rather than the earth, is the hub of our planetary system:

> He, who through vast immensity can pierce,
> See worlds on worlds compose one universe,
> Observe how system into system runs.
> What other planets circle other suns,
> What varied Being peoples ev'ry star,
> May tell why Heav'n has made us as we are.
> (Alexander Pope, An Essay on Man, 1753)

The Age of the Universe and the Earth

Astrophysicists have determined the age of the universe. One method of measuring the age of galaxies has relied on measurements of the distance and velocity of a receding galaxy. For speeding galactic material to have moved so far it must have started its lengthy journey 14 billion years ago—the universe is that old.

On the earth the accurate dating of geological and biological materials employs radiometric dating that capitalizes on the natural instability of some atomic nuclei. Simple molecules like salt and sugar are composed of some two to twenty atoms of various sizes. The simplest atom is hydrogen. It nucleus contains one positively charged proton with one negatively charged electron swirling around the nucleus (not to be confused with the nucleus of a cell, which is also at the center of things). The nuclei of all other atoms contain a mixture of protons (positive charge) and neutrons (a proton-electron merger that has no charge). Some types of atoms have unstable nuclei which occasionally break down and release particles or energy in the process.

What does radiometric dating tell us about the earth's age? Geologists can determine the age of Earth's most ancient volcanic rocks by measuring some radioactive elements that decay very slowly. Uranium-235 decays to an unusual form of lead: lead-207. (The atomic number 235 is the sum of the 92 protons and 143 neutrons in a uranium atom.) If uranium-235 and lead-207 are present in equal amounts, it would mean the rock was 713 million years old since that is the time required for half of any uranium-235 to decay, leaving as products 50% lead and 50% uranium. In another system of radioactive decay it takes 1.25 billion years for half of potassium-40 to decay to argon-40. By using these two methods along with confirmation from about 40 additional systems of radioactive decay, geologists have dated the age of our planet by comparing the proportions of radioisotopes in hundreds of thousands of rock samples evaluated by thousands of laboratories. Christian and non-Christian geologists agree that the oldest rocks on Earth formed 4.6 billion years ago ±0.5 billion years. This consistent value is independent of the assumptions specific for a given system of radioactive decay.

The age of an organism is indicated by the age of the stone in which the fossil specimen is embedded. The oldest cells lie in rocks about 3.5 billion years old. Numerous fossils of complex organisms are embedded in rocks that are between 0.5 to 1 billion years old. If sediment is rich in fossils, it typically indicates that organisms died, settled onto the bed of the sea floor, and became fossils when their preserved bodies were covered with silt and slowly replaced by minerals. The passage of millions of years allows successive sedimentary layers to reveal the relative ages of fossils embedded in each layer. Occasionally, animals were preserved without mineralization. For example, a hunk of ancient amber or pine tar may contain wholly intact insects, like an ant or aphid exquisitely preserved after some 40 million years. The message is clear; planet Earth formed more than 4 billion years ago and life on it has evolved over much of this time both in the sea and on the land.

The Concept of Plate Tectonics

From the perspective of our brief human lives we commonly assume that the earth has always been much like it is at the moment. Except for wind-driven waves and tidal surges, each sea seems confined to its basin; the climate appears to be rather stable; and the mountain ranges rigid and unyielding—at least in our personal experience. Yet, the surface of our planet is actually made up of about a dozen floating tectonic plates, each bigger than a continent. These massive plates drift slowly, moving perhaps one inch a year. It is the slowness of tectonic movements that permits an illusion of geological permanence. Over deep geological time some continental landmasses have pulled apart, allowing oceans of water to gurgle into the gaps of separation. Elsewhere, mountain ranges rose up as the edges of colliding plates buckled under unimaginably powerful forces. Relentlessly sliding and grinding, tectonic changes have been at work for more than 200 million years shifting the oceans and spreading out the continents. If we search for major climate changes in this mix, we discover prolonged ice ages whose massive glaciers scoured out pits that filled with melt-water that we may enjoy today as glacial lakes. Geologists predict marked changes in the shape of continents in the next 10 million years, a geological eye-blink at less than 0.2% of the age of the earth. Los Angeles will move so far north that it will bump against San Francisco. Somewhat later, in its northward drift, Africa will collide with southern France and Spain generating a powerful upward thrust that produces a mountain range to replace the by then dried-up Mediterranean Sea. The only constant in geology and biology is the inevitability of change.

Radiometric Dating of More Recent Organisms

The great majority of the world's carbon atoms have a nucleus with six protons and six neutrons, which explains why it is called carbon-12. Occasional carbon atoms have two extra neutrons, providing a total of six protons and eight neutrons in carbon-14. Now the interesting thing about carbon-14 is that its unstable nucleus makes it radioactive. Every so often a neutron releases a bound electron, leaving behind a proton. The energy of the electron escaping from the nucleus can be detected as relatively weak radiation. This nuclear decay converts an atom of carbon-14 (6 protons and 8 neutrons) into nitrogen-14 (7 protons and 7 neutrons). How often does this nuclear decay occur? If you had a fresh skeletal bone containing C-14, there would be half as much C-14 if it were re-examined 5,730 years from now. So with its half-life of 5,730 years, carbon-14 will decay to 25% of its original concentration in 11,460 years (5,730 x 2 years). Expectations of the starting amount of C-14 contained in the tissues of an organism when it was alive allow estimates of the age of any plant or animal that has been dead for less than about 50,000 years. Imagine that tissue from a woolly mammoth, found frozen solid in Siberia, had only 12.5% of the expected C-14. This would indicate the mammoth lived 5,730 x 3 or 17,190 years ago.

Roughing it on Earth—Extinctions Galore

This Earth, this massive, rock-encrusted planet we call home, congealed into a sphere 4.5 billion years ago only to endure repeated barrages of thousands of giant meteors pummeling its hot, molten surface. One massive collision may have flung off a huge glob of material that cooled to become our moon. Roughly 3.9 billion years ago the earth's surface had cooled enough to crust over. Enormous amounts of water vapor, carbon dioxide and nitrogen and sulfur gases were released by the violent eruptions of numerous active volcanoes. Even though a massive amount of water vapor was released, water could not condense into lakes or oceans because the earth was so hot that water only existed as steam. Yet, life's emergence seems to have wasted little time in geological terms, for there is fossil evidence of filamentous bacteria as old as 3.5 billion years. Today, in some saltwater lagoons in remote northwestern Australia you can see large mats of living bacteria similar to these ancient fossils. Bacteria are already complex organisms. So, before bacteria arose there must have been even less complex life in the form of replicating macromolecules that may have evolved soon after the earth crusted over.

As they examine organisms on Earth, biologists find delight in life's complex inner workings and its multitude of exquisite colors and forms. Most of us, caught up in our own daily routines, would not be inclined to pause and think about some earlier creature, a former life lying silent and undisturbed, embedded in ancient layers of soil beneath our scurrying feet. Nevertheless, the exposure of such fossils, those mineralized remains of preserved carcasses, reveals much about the history of life on Earth. Whereas humans reverently bury their dead in tidy graveyards, most dying organisms merely sink where they fall in the water or soil, only rarely fossilizing for posterity. The scientist's exhilaration at unearthing a fossil as dramatic as a walking whale or a flying reptile is tempered by the sad realization that none of these creatures is alive today.

Biologists also grieve for the irretrievable loss of creatures like the large Tasmanian wolves, many flightless birds, and delicately colored butterflies extinguished by the callous hands of man. Environmental abuse and overuse are now threatening blue parrots, Bengal tigers, chimpanzees, and thousands of other splendid and remarkable creatures. We hold the life of each species to be beautiful and precious, for to a human heart extinction seems such a tragic loss. Biologists, with more sorrow than rationality, deeply mourn the extinction of any living organism, however lowly its life. Evolution proceeds both through local competitions and major catastrophes. Cycles of global freezing and heating and collisions with asteroids have been responsible for some major extinctions. Yet, even had there been no ice ages, volcanic eruptions, nor other colossal catastrophes, there still would have been frequent extinctions as more successful competitors won out in the survival of the fittest. Ninety-nine percent of the species that ever flew, swam, or scampered about on our planet are now extinct. Genetic evolutionary changes require repeated cycles of birth and death to select for the best-suited mutations. Evolution can be faster in species with short lifespans that quicken the pace of successive generations. Each of today's living organisms reflects the accumulation of positive mutations whose utility makes it likely these genetic alterations will be passed down through future generations.

Photosynthesis is one of a handful of innovations now considered absolutely essential for the evolution of advanced forms of life. Photosynthesis is the process that uses the energy of sunlight to combine carbon dioxide and water into the sugar, glucose. Humans breathe the oxygen that photosynthesis produces as a by-product. Cyanobacteria, whose descendants remain abundant today, may have been some of the first cells to produce oxygen by photosynthesis more than 2.7 billion years ago. Before the emergence of cyanobacteria there were only traces of oxygen in the primitive atmosphere, which was probably just as well. Oxygen reacts so readily with other chemicals it might have prevented the formation of some simple molecules important for the initiation of life.

Evidence from fossils and from chemistry indicates that as far back as 2.1 billion years ago some cells had a nucleus. Nucleated cells were in for some rough sledding along the historical trail of life. Between 570 and 750 million years ago the earth spent millions of years as a giant ice ball. These freeze-ups were so intense that glaciers advanced all the way to the equator while the oceans froze to a depth of half a mile because the average temperature was 60°F below zero! Yet, some nucleated cells managed to persist near hot water vents or even in the ice. Slowly the earth warmed up again, perhaps because the sun grew hotter, or the earth tilted even more toward the sun, or an increase in atmospheric carbon dioxide gas trapped heat and warmed the earth as it is doing excessively today. After thawing out from these great freezes, descendants of surviving cells evolved into multicellular organisms. From these the "Cambrian Explosion" of 500 million years ago produced richly diverse organisms: starfish, clams, worms, shrimp and many now extinct, bizarrely shaped creatures. Seventy million years ago the top predators stalking the earth were steely-eyed dinosaurs. Hidden in niches beneath them were a few small mammals. Presumably held in check by predatory reptiles, our most distant mammalian ancestors found refuges in the sea or in the ground. The situation changed 65 million years ago when the colossal Chicxulub (Chick'-shoe-lub) meteor slammed into Mexico's Yucatan peninsula. After billowing dust and virtually immediate darkness, long term ecological disturbances and roiling climate changes apparently devastated the egg-laying dinosaurs. What is certain is that graceful, fearsome, and frequently stupendous dinosaurs died out within 300,000 years of this collision. New data has quashed the traditional scientific view that this extinction of the dinosaurs cleared the way for the rapid diversification of mammals. A recent comprehensive study of fossils and molecular data from virtually all existing types of mammals (4510 of 4554 species) shows that many types of mammals were already present some 20 million years before the disappearance of dinosaurs (Bininda-Emonds et al., 2007). No less interesting, the broad radiation into an assortment of mammals did not even start until 10 million years after the dinosaurs died out. Apparently it is one thing to have niches open up; it may be quite another for natural selection to fill them. But let's go back to the fascinating question of the origins of life on Earth.

Life on Earth Began Some 3.5 Billion Years Ago

The early synthesis of organic chemicals. If we can understand in more detail where and how life first arose and what form it took, it will help us estimate the likelihood that life has evolved on other planets in our galaxy. Perhaps planet Earth had some favorable spots or niches located along the edge of the lapping seas where cyclical exposure to ocean water, the atmosphere, sunlight, and drying on clay could unite chemicals into especially interesting molecules. Another possible site for the synthesis of interesting organic molecules is close to the mouth of superheated thermal vents in the ocean. Even today these remain as "hot spots" for novel creatures. It is extraordinary that some bacteria live "happily" at 121°C, considering that water boils at 100°C (Cowen, 2004). The early oceans of 2.0 to 3.5 billion years ago may have been as warm as 60-80°C (Knauth, 2005). The combination of energy from heat and the outward venting of nutrients may create interesting molecules. Organic molecules may also have cruised in on meteors or been mixed in with the dust of asteroids. Conceivably these sources contributed useful chemicals. Of course, if some form of life has been seeded on Earth from an extraterrestrial source, one would still need to answer the same basic question, "How did life get started at some other place in the universe?" Wherever life began, its constituent molecules must have been relatively simple at the outset. Some of the complex chemicals which today's living organisms use in synthesizing proteins are too recently evolved to have contributed to the origins of life (Wolf and Koonin, 2007).

The molecules of living creatures are composed of remarkably few elements; the most important are carbon, hydrogen, nitrogen, oxygen, and phosphorous. The challenge is to understand how complex molecules were formed and how repeated copying perpetuated life. It will be particularly important to know how proteins were made. Proteins are among the most interesting molecules in our body. While muscles and tendons have abundant structural proteins, most other proteins have chemical functions. In their diverse roles as chemical artisans, proteins split or unite other molecules, move molecules or ions, bind with select target molecules, regulate the activity of genes, accelerate chemical reactions, and detect various sensory stimuli like heat and light. We should praise those proteins whose movements contract our hearts and move our limbs, and those that extrude waste products, protect the body surface, carry oxygen from the lungs into our tissues and transport nutrients from our gut into our body, synthesize molecules, contribute to the function of brain cells, and most relevant for our immediate concerns, attach to other proteins that regulate gene activity (gene expression). Given their diverse and indispensable roles there is reason to honor proteins as molecules for all seasons.

Every protein is composed of chains of amino acids. Glycine is the simplest of the 20 or so types of amino acids. It is built up from a combination of four kinds of atoms: two oxygens (O), two carbons (C), five hydrogens (H), and one nitrogen (N). When proteins mature, their strands of amino acids (Figure 2.1) are likely to twist and fold into sheets or globs.

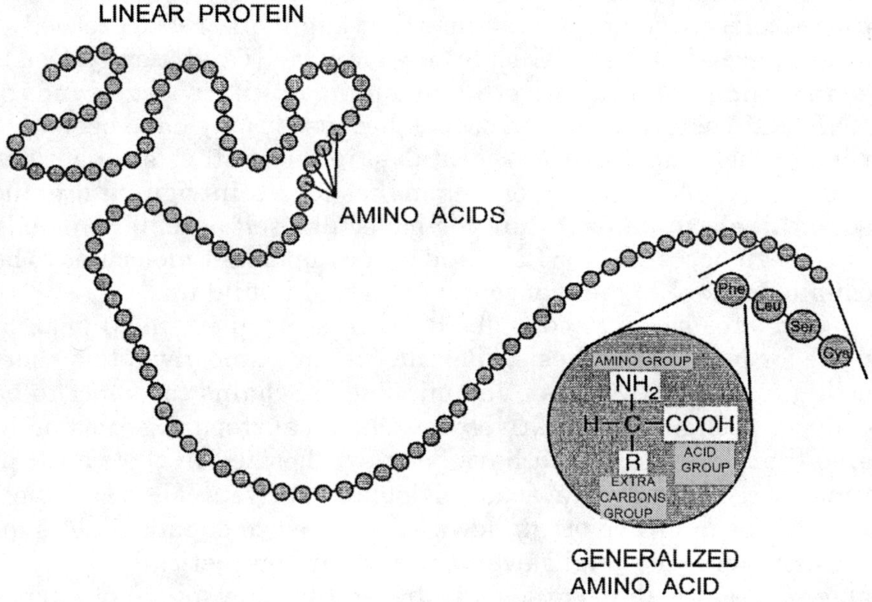

Figure 2.1 The primary structure of a protein as a linear chain of amino acids. Abbreviations indicate the four amino acids Phenylalanine, Leucine, Serine, and Cysteine. As shown, every amino acid has an amino group (NH_2) and at least one carboxyl acid group (COOH). The 20 or so naturally occurring amino acids all fit the general formula where R represents a hydrogen atom in the case of the smallest amino acid, glycine, or more generally a chain of one or more carbons and associated atoms for the larger amino acids. See Figure 14.1 for a ribbon representation of a coiled and folded protein.

The original synthesis of proteins and other complex organic molecules. Frustratingly little is understood about the origins of life on

Earth. As a result there are multiple scientific hypotheses about how life got its start. To provide a sense of thinking about life's origins we can focus on one popular scenario: an important role for RNA (ribonucleic acid) in the emergence of life. Many interesting molecules seem to have formed in the early chemical conditions on Earth when there was much more carbon, nitrogen, and hydrogen than oxygen. All of the important amino acids, all five of the nucleotides in DNA and RNA (see below), and several lipids (fats and oils) and other organic molecules have been formed in laboratory simulations of the early chemical environment. On primitive Earth inorganic catalysts may have helped to synthesize higher concentrations of important molecules. (Catalysts are molecules that accelerate chemical reactions that might otherwise proceed at a slow pace.) Patches of silicate clay may have catalytically assisted in joining chemicals together including uniting short strands of RNA. RNA could have been crucial because it may have been able to copy itself and also act as a catalyst in the synthesis of peptides and other molecules. All of this makes for an intriguing organic soup. Likely to be too dilute for molecular self-assembly, it may have been necessary for RNA and other important molecules to be confined where higher concentrations could build up.

It makes sense to consider RNA as a primitive mold or template for peptide synthesis. "Peptide" is the name given to a short string of two to twenty or so amino acids, chains too short to be called proteins. With relatively few chemical steps, it seems probable that peptides would have been synthesized in the course of time. Large peptides can act as biological catalysts called enzymes. Enzymes can either break down or synthesize chemicals. A synthesizing enzyme might even catalyze an energetically unfavorable reaction if the synthesis is driven by some source of energy like supplemental heat, light, UV radiation, electricity, pressure, or ionic concentration gradients.

What seems apparent is that fairly early in the evolution of life, chains of RNA arose that used an early version of today's code for combining amino acid building blocks into peptides. Such early codes—protocodes—have now disappeared from view. RNA, and later on, DNA (deoxyribonucleic acid), would eventually provide a standardized alphabet. As we will see, there exists a remarkable universality in the genetic instructions of all forms of life on Earth. However it came about, both the copying or replication of molecules, and the selection for still more effective replicating molecules were events at the core of life's beginnings.

Capturing energy. Living organisms are not closed systems. Like a car that needs gasoline to run, organisms also need to have some external energy source for their cells to run. Even elemental life must somehow have captured enough energy, perhaps obtained from iron or other minerals that bind oxygen (oxidation), or by having sunlight drive the synthesis of glucose that could later be oxidized for energy. In dealing with today's creatures in habitats like tropical forests, oceans, or grassy savannas, we can identify "food webs" where the raw tissues of other animals and plants supply needed energy. But the earliest life, replicating molecules, would have needed other sources of energy because eating, in the ordinary sense, was not a possibility at the outset.

Identifying early mechanisms of natural selection. One of the important challenges is to find principles of natural selection that were effective at the onset of life. Some of the difficulties in understanding the early evolution of life are highlighted by the miss-phrased question: How could several species of RNA engage in a "struggle for survival of the fittest"? What could these molecules be struggling for, how could the outcome of the struggle favor the survival of some, and in what way would one RNA molecule be more fit than another? The "struggle for survival of the fittest" is an evolutionary phrase better suited for describing the struggles among competing organisms than among molecules. At the level of molecules, it is more accurate to think of types of self-replicating catalysts "competing" for chemicals to be used for synthesis; only figuratively do they eat chemical food. The catalytic chemical that is best at using up other nearby chemicals may have had a selective advantage. Perhaps a particular variant of catalytic RNA, a ribozyme, was the fittest because it was best able to catalyze the synthesis of molecules, notably itself, when powered by energy it gained in using the products of other less reactive ribozymes. One would expect that an RNA molecule that was stable and rapid in replication would be a winner among a variety of RNA molecules. You should note that this scenario is speculative; there is no settled understanding of the emergence of early replicator molecules.

Tracing out the probable path of life's early evolution is a daunting task. Even finding a feasible path is a challenge, but perhaps not an insurmountable one. To that end, Harvard University will spend one million dollars a year on their "Origins of Life in the Universe" initiative. It will be interesting to consider the models they formulate for a likely series of steps leading from inanimate chemicals to primal life. No scientist will conjure up vitalism or the intrusion of miraculous forces as a necessary fac-

tor.
Defining life. What are the universal properties of life on Earth? At a minimum, a living organism should display all of these five features:

1. DNA- or RNA-based genes whose protein products guide development and growth. (Even *E. coli* and other bacteria have more than 4,000 genes, so they are hardly simple).
2. Consumption of energy from the environment—life is not a closed system.
3. Responses to environmental properties or changes.
4. Replication or reproduction.
5. A capacity to evolve by incorporating heritable genetic changes.

Note that neither a self-replicating mechanical robot that assembles components from a parts bin nor a self-replicating computer program would meet the first requirement for life. The invention of a DNA-based, self-replicating robot is the stuff of science fiction novels—at least for now.

The world has many self-assembling systems that are not alive. By themselves physical forces and chemical bonding have generated a multitude of complex regularities, like the stalagmites and stalactites that elongate in caves. Some of these fragile mineral stalactites are as thin as soda straws and more than 15 feet long. Crystals of colorful minerals like iron pyrites, or fool's gold, come in many dazzling geometric forms. Snowflakes have both regular patterns and individuality; no two are alike. These are interesting examples of ordered and individualized growth, but they are hardly alive.
The evolution of cells. The advent of the first primitive cell was a milestone event. Even if we knew those historic steps in the evolution of cells, it would be virtually impossible to mimic them today. Like waiting for rain to fall on the world's most arid desert in Chile, the desired chemical reaction might eventually happen without assistance, but probably not for centuries or longer. While a scientist can only devote a few years to a project, in the natural world the thousands upon thousands of years available for actual evolution make low probability events feasible. Early protocells were surely much less complex than modern cells; their outer membrane was probably a simple lipid bilayer. Because oil (lipid) and water do not mix, lipids have a natural tendency to form spherical bilayers that could enclose a starter mix of molecu-

lar soup. The "water-hating" portions of proteins would readily burrow into the lipid bilayer. The natural emergence of a protocell over time is certainly plausible. Even today, the external membrane enclosing a typical animal cell is a bilayer of lipids with a smattering of proteins adrift in it. Energy storage and competitive success have recently been demonstrated in more elaborate models of protocells (Chen, 2006).

Of course, it would be especially interesting to devise ways to construct a "protocell" from off-the-shelf, laboratory components, such as enzymes and a few molecular companions—all enclosed in a membrane. It is already possible to transplant bacterial genes to convert one bacterial species into another (Lartigue et al., 2007). Biologists assume the first elementary cells had the simplest of features, in sharp contrast to today's highly evolved cells which are bustling chemical communities. The multiple varieties of human cells, for example, deploy internal chemical messengers able to act through more than 500 different kinases; enzymes specialized to control the functions of proteins by activating them with phosphate groups. Enclosed by a protective membrane, modern cells are stuffed with structural proteins for struts and functional proteins for enzymes, and receptors that bind select chemicals, and pumps that move and concentrate chemicals. Cells contain numerous membranous organelles whose sacs hold various chemicals. The mitochondria and chloroplasts within cells are remnants of ancient hitchhikers that themselves were once primitive cells. Uninvited, they long ago invaded larger cells and have overstayed their visit by a billion years or so. All animal cells have mitochondria; plant cells have both mitochondria and chloroplasts (Figure 2.2).

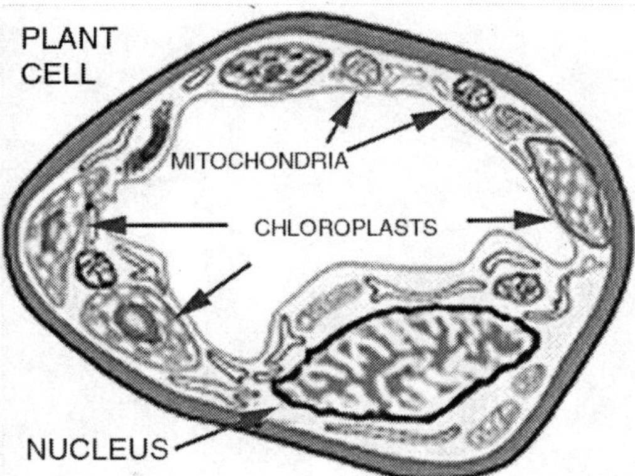

Figure 2.2 This plant cells contains a number of organelles, including

the nucleus and some mitochondria and chloroplasts, as labeled. The nucleus is packed with DNA. Mitochondria are concerned with energy production. The chloroplast uses the sun's radiation in the synthesis of glucose. The very large white space in the center is a fluid-filled vacuole, a cavity characteristic of plant cells.

After life began, it thrived in the water and also clung tenaciously to the soil of this remarkable planet. The development of primitive cells, the production of high-energy chemical bonds, the biochemical use of sunlight to make sugar with oxygen as a by-product, and the onset of multicellularity are some of the high-points in the early evolution of life. Prolonged ice ages, occasional collisions with huge asteroids, or other apocalyptic mega-events have wiped out whole groups of plants and animals, but diverse forms of life have persisted. In contemplating the evolution of life on Earth, we surely owe some of the world's smallest creatures an enormous debt, especially the cyanobacteria and similar tiny oceanic plants that release the oxygen we now breathe. Hail to this photosynthetic plankton and to green plants. Even some physical factors warrant appreciation. It was only after the formation of oceans that our early ancestors had opportunities to evolve in the sea and ultimately emerge to colonize the land. Then, as now, life was based upon the chemistry of seawater. The salts that surround human cells strongly resemble dilute seawater that contains sodium, potassium, phosphorous, chloride, and other useful minerals. So much to be grateful for about life: how it arose and flourished in many wonderful ways.

3

&

The Universal Code of Life

DNA: A Code For All Seasons

The Christian Bible, the Qur'an (Koran), and the Hebrew Bible were all compiled long before the discovery of the DNA Code of Life. Although humans have always been surrounded by an abundance of messages from the Code of Life, as splendidly translated into plants and animals, all of its billions of individual texts have been completely hidden from view. The informational function of DNA was so well-disguised that until the mid-1940s few people suspected that DNA had a pivotal role in life. Moreover, although life's language uses only 64 arrangements (four types of letters taken three at a time), this linguistic code of DNA was not broken until the 1960s. It has only been in the past few years that the enormously long strings of four letters—the DNA text or genome—was completely read for any organism. Even now delightful poetic accounts of human origins remain far more emotionally evocative than the numbingly long sequences of seemingly jumbled letters whose 64 triplet combinations encode in exquisite detail how every organism develops. But the truth of the DNA code has been proven by experimental analyses and by satisfying practical applications in biology and medicine. The genetic revelations embedded in the Code of Life are replacing all other creation stories about life on Earth. To acquire the exact meaning of each organism's personal text requires a precise translation of its lengthy sequence of DNA code. Since the DNA code is a universal code, understanding how expression leads to a protein applies in general terms to any organism. Those encountering these concepts for the first time should carefully examine the figures to grasp the meaning of the several new terms. For those wishing somewhat more detail, the appropriate chapters in a textbook of biology would provide a helpful introduction.

Molecular Genetics: The Fundamentals of Life's Code

Deoxyribonucleic acid. The DNA of multicellular organisms is coiled and compacted into linear chromosomes visible during certain stages of cell division. When chromosomal DNA is stretched way out, it becomes evident that the DNA consists of two aligned strands. In its native state these two weakly bound strands of DNA are twisted in a tight spiral. One can imagine this double-stranded DNA as an extremely long, twisted rope ladder in which each rung is one letter of the code. Many biologists consider that figuring out the ladder-like structure of DNA was the greatest biological discovery of the 20th century. It earned James Watson and Francis Crick the Nobel Prize in 1962 (Figure 3.1).

Nucleotides and bases. Originating billions of years ago, the world's most ancient scripture, DNA, can be translated from long sequences of the bases adenine (A), thymine (T), guanine (G), and cytosine (C). These four compounds have historically been termed bases because they are somewhat basic or alkaline rather than acidic. A strand of DNA is a long string of nucleotides, each composed of a sugar and its laterally extending base. Thymine and cytosine in DNA have a simple structure. Each is a ring made out of four carbon atoms plus two nitrogen atoms decorated with hydrogen or oxygen. The other two bases, adenine and guanine, are double rings. A nucleotide consists of one of these four bases attached to phosphate and the sugar deoxyribose. DNA consists of long strings of these four nucleotides, ATGC, in various orders and quantities. Imagine thousands of beads in only four colors strung on a very long string in many color combinations. The four bases in RNA are the same as the four bases in DNA except that the base uracil (U) is substituted for thymine (T) (Figure 3.2).

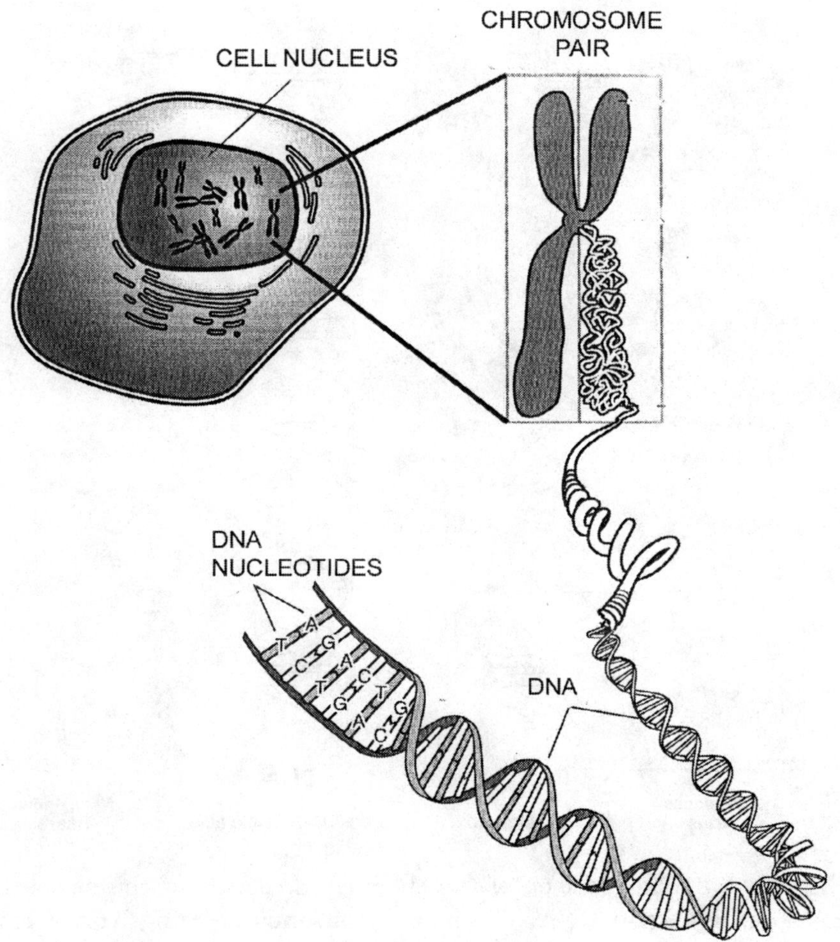

Figure 3.1 The nucleus contains the cell's complete set of paired chromosomes. An expansion of one chromosomal pair, say chromosome 15 in humans, shows an attachment point near the center of each chromosome. Stretching out the DNA reveals that its two spiraling strands are bound together by rungs of nucleotides.

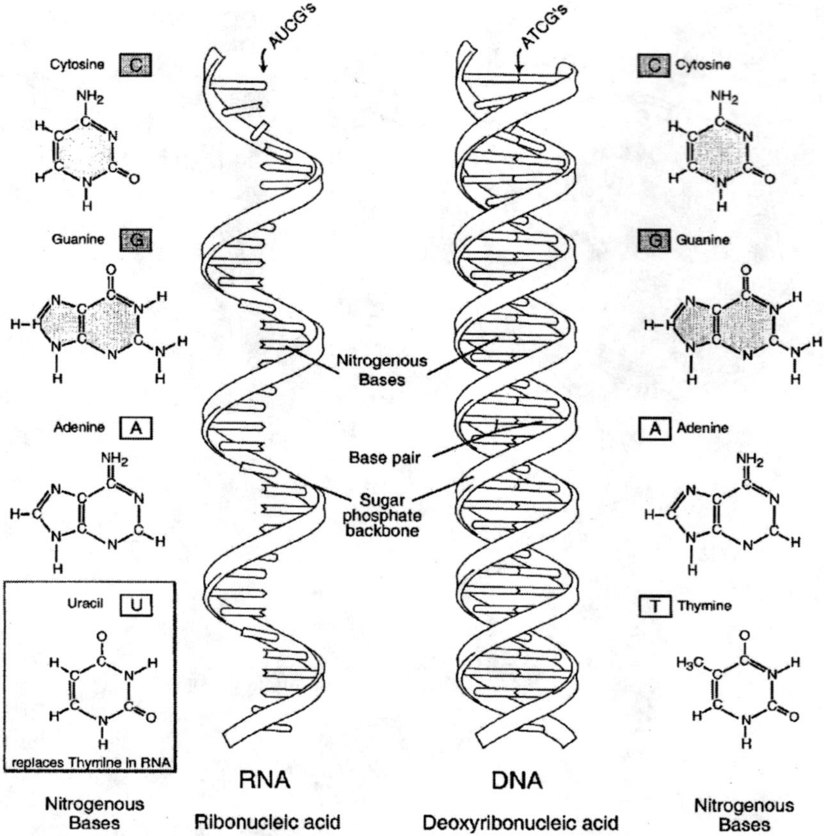

Figure 3.2 Comparison of the chemical structure of the bases that make up RNA and DNA. The many thousands of "rungs" are formed from complementary pairs of bases—in every strand of DNA, G and C join together as one rung or A and T as the alternative rung. In RNA U replaces each T. See the box.

When we look in still more detail at the bonds between RNA and a complementary strand of DNA we can see that the bases project off to the side of the sugar (deoxyribose) backbone. This permits the corresponding bases on the two strands to approach closely enough for adjacent hydrogen atoms to bond together. Adenine and Thymine (AT) always pair together as do Guanine and Cytosine (GC), as shown in Figure 3.3 by their convenient first letter abbreviations. Uracil in RNA always binds to Adenine (UA).

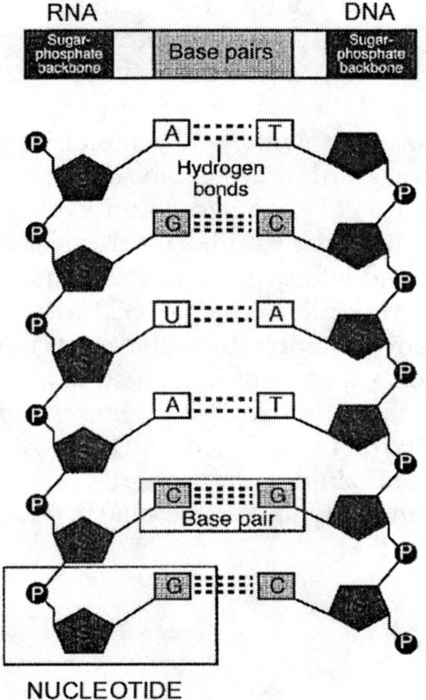

Figure 3.3 Base pairing of RNA with DNA. The two columns of black pentagonal (five-sided) figures represent the sugar ribose for RNA or deoxyribose for DNA. Phosphate (P) links these sugars together. Note the extra hydrogen bonds joining the CG and GC base pairs.

The DNA double helix can be unwound and separated into two single strands. Single-stranded DNA will readily bind to a complementary strand of RNA. To give a feeling for the code, let's figure out what sequence of three RNA bases would bind to the triplet TCA in a strand of DNA. Can you visualize that AGU on the RNA would bind to the complementary TCA on the DNA strand and build the three rungs AT, GC, and UA? (See the top half of Figure 3.3 for precisely this pairing.) For both DNA and RNA the exact order of bases also matters for coding. Hence, the sequences AGC and CGA are different codes.

How does the base pair code work? The shortest coding sequence consists of three letters; it represents one amino acid. For example, the triplet AAG in RNA codes for lysine, one of several amino acids we must have in our diets in order to survive. The use of a three-letter code for amino acids is characteristic of all living organisms from microbes to man. The elegant simplicity of this code and its universality inspire a reverence for life, especially in those who enjoy the wonderful privilege of decoding the genes. The Code of Life warrants a prominent place among revered texts. It has extensively field-tested instructions that direct life on Earth. Close similarities of DNA base sequences reveal the ancestral relatedness of organisms and even the paths taken by our ancestors in their evolution. The translation of DNA reveals secrets to health and longevity, if not a fountain of youth.

Having four available bases (ACGT) allows 64 different sets of three bases, more than enough distinct triplets to code for the 20 or so amino acids that organisms use to build proteins. In fact most amino acids are represented by more than one triplet code, for example, the triplets GGU, GGC, GGA, and GGG all code for glycine. Rather amazingly, the actual Code of Life, the triplet codes for all 20 amino acids, can be neatly summarized in a compact table (Figure 3.4).

SECOND LETTER

FIRST LETTER	U	C	A	G	THIRD LETTER
U	UUU UUC } Phe UUA UUG } Leu	UCU UCC UCA UCG } Ser	UAU UAC } Tyr UAA Stop UAG Stop	UGU UGC } Cys UGA Stop UGG Trp	U C A G
C	CUU CUC CUA CUG } Leu	CCU CCC CCA CCG } Pro	CAU CAC } His CAA CAG } Gln	CGU CGC CGA CGG } Arg	U C A G
A	AUU AUC AUA } Ile AUG Met	ACU ACC ACA ACG } Thr	AAU AAC } Asn AAA AAG } Lys	AGU AGC } Ser AGA AGG } Arg	U C A G
G	GUU GUC GUA GUG } Val	GCU GCC GCA GCG } Ala	GAU GAC } Asp GAA GAG } Glu	GGU GGC GGA GGG } Gly	U C A G

Figure 3.4 The Universal Genetic Code. The code for each amino acid is presented as a string of three RNA nucleotides. As shown by the small shaded box, the second row (C) intersects with the fourth column to provide four triplet codes that all begin with CG. The third letter can be disregarded for some amino acids. For example, the combinations CGU, CGC, CGA, and CGG all code for arginine. This shows that the first two letters are the most important in the genetic code. UAA and UAG are codes that act to halt or stop further reading along the nucleotide string of RNA. The amino acid abbreviations represent: phenylalanine, leucine, isoleucine, methionine, valine, serine, proline, threonine, alanine, tyrosine, histidine, glutamine, asparagine, lysine, aspartic acid, glutamic acid, cysteine, tryptophan, arginine, and glycine

Traditionally, a gene is a long sequence of triplets that codes for the series of amino acids that will make up a particular protein. In the simplest case one gene codes for one protein. The protein insulin, which helps some cells take up the sugar glucose for storage into starch, is made up of 51 amino acids encoded by a strand of 51 triplets or 51 x 3 bases of DNA. As a counterpart to insulin, glucagon is a smaller protein composed of 29 amino acids that helps to convert starch back into glucose for energy.

Humans have some 25,000 genes dispersed among their 23

pairs of chromosomes, or about 1,100 genes per chromosome. These genes are quite scattered within each chromosome; they represent perhaps less than 1% of the nucleotides in each chromosome's very long string of nucleotides.

There are two fundamental steps in the read-out of a gene leading to protein synthesis. The first step is the absolutely critical synthesis of RNA (transcription) from a template of DNA (the gene). This happens in the cell's nucleus where an unwound portion of DNA is transcribed into a corresponding strand of RNA. Upon leaving the nucleus, this messenger RNA (mRNA) serves in turn as a template for the translation of its code into a string of amino acids. While the mRNA is being read triplet-by-triplet, short pieces of RNA, called transfer RNA, ferry in specific individual amino acids that add to and lengthen the developing protein. A transfer RNA will only release its particular amino acid if its own triplet code matches up with the corresponding three-letter code on the next portion of mRNA to be read out (Figure 3.5).

Let's assume that from DNA a very short messenger RNA has been made whose coding region includes the base sequence UUCGAC. UUC codes for the amino acid phenylalanine and GAC for the amino acid aspartic acid. The translation of the UUCGAC code will join these two amino acids into a dipeptide. In a commercial application a slightly modified version of this dipeptide is the artificial sweetener aspartame. You may recognize aspartame as the billion dollar commercial product Nutrasweet® that sweetens much of today's diet soda pop.

Of course, to make even an average-sized protein would require a much longer sequence of messenger RNA. For example, to make a protein with 100 amino acids would require translation of a precisely ordered RNA sequence of 100 triplets or 300 bases. In this manner the 100 amino acid building blocks are linked together in exactly the right sequence. A cell must accurately repeat this synthesis many, many times to produce useful amounts of the protein since a solitary protein molecule isn't useful.

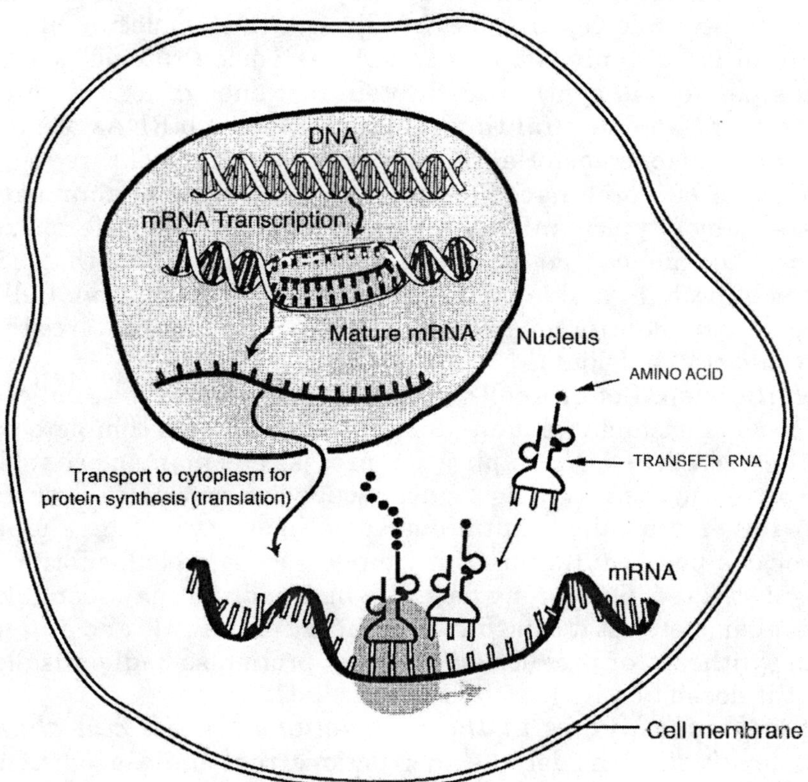

Figure 3.5 This drawing of a cross-section through a cell reveals the formation of a strand of RNA being transcribed from an unwound stretch of DNA. Following some trimming, the mature messenger RNA leaves the nucleus to serve as a template for protein synthesis. A transfer RNA that matches up with the next three bases on RNA will add its amino acid to the elongating protein. The protein in the process of being synthesized in this example already has a chain of eight amino acids linked together.

Cells are the basic units that make up living creatures. Although cells are complicated, the fundamentals are known—there is no evidence cells are infused and empowered by some mysterious vitalism. Consider protein synthesis in bacteria. The best studied of all bacteria is *E. coli* like those living in our gut. We don't have to review the multiple roles of proteins in order to get a feeling for the scope of protein synthesis. Cells like *E. coli* are living factories that decode messenger RNAs to make proteins.

Because bacterial mRNAs are rapidly broken soon after use, they need to be made continuously. A given mRNA molecule may remain intact for only two minutes. An extension rate of 15 amino acids per second synthesizes protein fast enough for the mRNA to be translated several times. Many different mRNAs are present. Overall, one bacterial cell contains an average of three copies of each of 600 protein-coding mRNAs. The string of amino acids encoded by a typical mRNA is broken up into three different proteins. This means that an *E. coli* cell synthesizes more than 1800 types of protein, making about 1,000 copies of each kind. Cell biologists can identify the roles of most bacterial proteins, like those that aid cell division.

Cell division. Before a cell can divide, its DNA must be duplicated to provide each of two daughter cells with its own complete copy of the DNA. The DNA duplication process has many mechanisms to correct errors and insure a nearly flawless copy. However, rare copying errors called mutations sometimes occur. The simplest error is a point mutation in which one base is added, subtracted, or substituted. One wrong base among hundreds may seem like a trivial matter, but it can shift the reading of the code and halt protein synthesis, or produce an aberrant protein so badly misfolded that it doesn't do its job.

Mutational changes in the instructions for life can change the product. Consider the instructions that build a particular car. They are selected from a menu of alternatives like one-tone or two-tone paint, dark or light upholstery, a cold-weather or a warm-weather thermostat, a standard or a turbo-charged engine, and so forth. The menu of genetic instructions for building multicellular organisms offers a similar choice between the two chromosomes in each pair. But there are interesting differences. There is an apparently random selection between the choice of either the mother's or the father's chromosome for each of the 23 chromosomes. More interestingly, the instructions may change if a piece of chromosome occupies a new position. As a result of such mutated instructions, the altered organism may be superior. Superior in the sense that the organism will produce more and healthier offspring than its forbearers did. Mutations turn out to be an effective way to ensure that every so often modified instructions will produce an organism that is better suited to its present environment. Not surprisingly, most genetic mutations are harmful. As a strategy, mutation works best across an extended period of time with many trials that provide multiple opportunities for the natural selection of a few useful mutations, such as fungus-resistant potatoes that spontaneously arose from a rare mutation among many generations of parental plants.

Box 3.1 RNA Trumps the Gene Concept.

The "central dogma" in the 1950s was that RNA never served as a template for DNA; the direction of synthesis was always from DNA to RNA to protein. But in the 1960s, enzymes were discovered that use RNA to make DNA (reverse transcription). Independently it was shown that many enzymes in single-celled organisms could slice DNA into specific small pieces. Harnessing these two discoveries made it possible to make and add pieces of DNA, including entire genes, into cells—genetic engineering was born. The recent determinations of the complete DNA sequence (millions of ATGC combinations) of dozens of plants and animals have spawned diverse advances in understanding. It is evident that portions of an individual vertebrate gene may normally be located in several places within and even between chromosomes. A remotely located piece of a gene might have additional functions according to its immediate surroundings.

Surprisingly only 1.5% of human genomic DNA represents identifiable classic genes. It was once thought that the remaining 98.5% of DNA was mere junk—remnant scraps left over from ancient times. Yet, it is now known that some 80% of all human DNA is transcribed into short and long pieces of RNA (Birney et al., 2007). Based upon the character of the RNA sequences and their short lengths, it is evident that most of human RNA does not code for protein, but instead regulates gene expression in multiple ways. Similarly, of the 66,000 types of RNA in mice, 34,000 do not code for protein. Such probable regulatory RNA is even more abundant in humans. Evidently, the workings of both mice and men are controlled by numerous short RNAs that influence the reading out of genes, including genes responsible for producing the body's protein parts (Buchan and Parker, 2007). Such a new view of the Code of Life shifts the emphasis away from DNA and from the classic gene concept to RNAs as extraordinarily important informational units with multiple influences on gene expression.

The Long and Short of Organismal Specializations

Whether a mutation improves the fitness of an organism depends not upon pure chance but upon the suitability or fitness of the change for its environmental context. Competitive rivalries among organisms along with the specific requirements of the environment set the context for evaluating the fitness of an organism with a novel mutation. Will the individual prosper or not? Will a mutant protein aid or impair its recipient's ability to leave numerous viable offspring? A valuable mutated gene may spread within the gene pool, i.e., spread among members of that species over many generations. Humans are like most other organisms in living fleeting, imperfect lives. The environmental context and the competition are keys to success and survival. Many poorly adapted species have become extinct—they came and went. Yet, even though species are ephemeral, many of their component genes can be nearly everlasting as they continue to spread and benefit modern species. Just as all brands of cars share numerous features like headlights and horns, so too thousands of living species share many genes through their own common lines of descent. Any two species that are closely alike probably branched off from a recent common ancestor analogous to the many versions of minivans founded on Chrysler's original innovation. Species that didn't thrive went extinct, like the Oldsmobile. Still, the utility of Oldsmobile's hydromatic shift insured that this invention would spread and carry on as the automatic transmission of other cars.

Ever since Darwin, studies of the Galapagos Islands indicate that finch beaks became specialized for each island's available food resources such as small insects or tiny seeds. Recently, severe drought changed the finches' food supply and revealed natural selection at work. The finches that fared best had beaks well-suited for cracking the heavy remaining seeds. A finch either had a strong beak, or it failed to eat. In this common situation involving rapid change over a span of only a few years, scientists were witnessing selection for beak strength from an existing pool of genetic variability, rather than selection for a brand-new mutation that happened to strengthen the beak. The mutations for genes responsible for stout beaks had already occurred long ago. This example about living island finches, along with much additional evidence, supports the view that, both in the long and short run, evolution proceeds toward specializations that are serviceable in the environment at hand.

Speciation

About 1.8 million living species have been described in some detail. Several million more have yet to be classified. Species form in various ways. Plant seeds may drift long distances in the wind, or animals may migrate overland and relocate in a geographically isolated valley. If small groups of these organisms remain isolated for a long time, it is likely that some genes will become more prevalent in subsequent generations (genetic drift). In addition, the combination of accumulated mutations and gene shuffling may produce offspring that would be unable to fertilize the evolved descendants of those organisms remaining at the original source. A new species has arisen.

Genetic similarity is an indication of shared ancestry. Comparing the similarity of genes across living species helps biologists to generate family trees. At the tree's base is the ancient root from which various groups of organisms branched out, splitting in an adaptive radiation of species over unimaginably long periods of time. Millions of years of separation have produced the enormous diversity we enjoy today like daffodils and daises, oaks and maples, dragonflies and butterflies, clams and oysters, sailfish and salmon, tortoises and hares. Yet, even after all of that diversification, profound genetic relatedness remains. One big happy family of life encoded by its DNA!

Remarkably, different species, even visibly different species, may share many of the same genes. Let us examine how even slightly different sets of developmental genes and regulatory sequences can produce a unique species.

Development

The embryo has the extraordinary challenge of developing into an adult. Biologists are coming to understand how this occurs. Development is a stringent filter that selects against most mutational change; the odds are high that any novel change will disrupt crucial networks of gene regulation and naturally abort the developmental process. Most mutant mammals fail to develop. Nonetheless, in the long run, some individuals will gain advantages from rare favorable mutations.

Body form. It seems like an insoluble problem to determine how genes produce body forms as different as those of a fly and a mouse. Then again, visibly dissimilar animals may have a deep and unexpected genetic relatedness. The paradoxical emergence of different forms from similar genes is being resolved by recent research in development that paints a vivid picture of the evolution of animal complexity. Gene expression is the actual use of RNA made from a gene's DNA template. At any given moment most genes are inactive, rather than being expressed. In some instances a gene is silenced by RNA or by a protein that binds regulatory sequences for that gene. Such a repressing protein may be attached to sequences of bases near the beginning of the gene. Until the repressor is removed, the gene cannot be read out—it cannot be copied into RNA. Even if a gene produces RNA, nearby enzymes may dice up and destroy the resulting mRNA. Developmental differences across organisms certainly reflect which genes are busy producing intact mRNA, in which cells these genes are functionally active along the length of the embryo, and, of course, when the genes are active during development.

Let's see how these principles are helpful in understanding the construction of the body of a fruit fly. The fly's body is made of a head-to-tail series of segments. It is a particularly telling finding that on chromosome number 3 the eight genes that control body segmentation (the *Hox* genes) lie in a tidy row in the same order as the sequence of head-to-tail body segments, head gene first—tail gene last. All eight *Hox* genes contain within them an identical sequence of 180 bases—the homeobox—in addition to longer sequences of code specific to each *Hox* gene. *Hox* genes and their homeobox domains were profound discoveries, because what is true for fruit flies is true for many other organisms, including humans. Like mice, fish, and flies, humans have *Hox* genes that line up in a head-to-tail order. Each *Hox* gene produces a protein that regulates the expression of genes important for body development. (The specific part of each Hox protein that physically binds and regulates the DNA is the cluster of 60 amino acids encoded by the sequence of 180 homeobox bases.) *Hox* genes are similar to one another because they arose from gene duplications that can be traced back to a founding *Hox* gene. Further analyses have made it clear that over a vast range of animals the forms of the body's major segments develop in head-to-tail succession because the *Hox* genes are read out in this succession. Mice have 38 and humans 39 *Hox* genes that regulate body form. Other developmentally important genes also reveal the relatedness of animals.

One such gene is *Pax 6* needed for the regulation of eye development. Humans and mice that lack *Pax 6* are eyeless. *Pax 6* expression can even activate its gene cluster in cells remote from the eye. Wherever a fly's local cells have been encouraged to express *Pax 6* an eye will form, even on the leg, or the wing, or on other tissues. In a remarkable experiment, the *Pax 6* gene of the fruit fly was replaced with the *Pax 6* gene from the mouse (Gehring, 2002). When the fly was examined it became clear that the mouse *Pax 6* gene had enabled the development of the normal faceted insect eye! Ninety percent of the amino acids are the same in the Pax 6 proteins of mouse and fly. Mouse and human Pax 6 proteins have the identical sequence of amino acids. Differences in gene mutations and duplications that regulate gene expression determine whether a mammalian eye or an insect eye develops. *Pax 6* is now recognized as a crucial gene that regulates eye development in humans, mice, fruit flies, squid, sea urchins, and other animals, including the eyespots in *Planaria*, a flatworm. It would appear that the most complex eyes arose after sequential modifications in a founding *Pax 6*-based genetic pathway that can be traced back to a spot of light-sensitive cells. Such changes may have happened several times.

Only a biologist would find the lithe "legs" of a sea urchin and an Irish river dancer similarly admirable. But guess what, like *Pax 6* for the eye, *Dll* is a crucial gene that helps to control the emergence of vertebrate appendages such as fins, wings, legs, and yes, even the "legs" of sea urchins. The sea urchin has emerged as a stunning source of many genes which until recently were believed to have evolved much later and exclusively within vertebrates (Samanta et al., 2006). As an echinoderm, the sea urchin may seem like a simple "pincushion" even more primitive than its more familiar relative, the starfish. Yet, it has 22,000 genes including genes that code for photoreceptor pigments and smell receptors and brain development much like those found in vertebrates. Evolution is a tale of emerging regulations of ancient genes as much as it is the appearance of new genes from recent mutations.

Working with yet another developmentally important gene, Rolf Bodmer and colleagues characterized a fruit fly gene which they called *Tinman*, after the Wizard of Oz character who lacked a heart. The Tinman protein activates one of several other genes required for heart development. Genetic cousins of the fly's *Tinman* gene are required for heart formation in mammals (Bodmer, 1993).

The wide distribution among animals of developmental genes such as *Hox*, *Pax 6*, *Dll*, and *Tinman-like* shows the relatedness of animals. What their proteins have in common is the ability to bind to regulatory sequences that control other genes responsible for crucial aspects of body form. A regulatory protein can either arouse or suppress gene expression. The sequence of bases in a typical structural gene might be preceded by a span of regulatory DNA having more than 10 sites where regulatory proteins can bind. This provides opportunities for several regulatory genes to act on one target gene. The fruit fly has about 13,700 genes and humans about 25,000, many of which are regulatory genes that control the expression of other genes.

In addition to providing a detailed set of instructions as a road map for development, such genes and their regulatory sequences are vast storehouses of history. Like the weathered face of a rugged fisherman that reflects a history of sometimes harrowing adventures at sea, so too the sequences of DNA nucleotides reveal the evolutionary history of an organism, specifically the mutational adventures of its ancestors. All of us have thousands of unheralded quite distant relatives who have provided copies of their genes that exist today as a part of our genetic instructions, often shared across diverse branches of the tree of life. There are numerous former lives to be grateful for.

Cellular differentiation is a fascinating developmental puzzle. How is it that the various cells descended from a fertilized human egg will in time differentiate in so many alternative ways; into nerve cells, heart muscle cells, bone cells, immune cells, skin, or hair? It is a puzzle because every one of these cells has the identical set of genes! Since DNA is duplicated prior to every cell division, each of these different types of cells carries the complete set of genes originally found in the fertilized egg. So the secret to getting skin rather than muscle is to arouse slightly different subsets of genes. Clearly, the key to cellular differentiation must be in the differential regulation of gene expression in particular cells at particular times. This is accomplished by chemical signals within and around the cell that control which genes are expressed and repressed (Carroll, 2005).

For the leopard to form one of its dark spots, a gene for dark pigment is activated in a patch of skin cells. Surrounding this patch is a doughnut-shaped ring of skin cells where the pigment gene is turned off by a repressor protein. In the skin region of overlap where repressor and activator proteins both bind to portions of a cell's pigment gene, repression (light color) wins out. In his manner, the activation of a dark pigment gene in a patch of skin cells and its repression in cells in the surround and the region of overlap give the leopard its dark spot(s).

The most critical questions to ask of the regulation of gene expression are: where? when? and for how long? Genes are expressed in some cells (where), beginning at exact times during development (when), for a length of time that controls the amount of protein produced (how long). Specific understanding of differentiation requires working out the regulatory details for each tissue.

Glimpsing genetic variation. There are about three billion bases in human chromosomal DNA. The initial work-up of the human genome suggested a low value of only 0.1% individual variation. This would mean you are about three million bases different from your neighbor. However, there is more variation in the human genome than was first appreciated from the initial analyses. For example, Redon et al. (2006) examined DNA from 270 people and showed that 12% of the nucleotides were deleted or duplicated. This genetic variation among different people is called copy number variation. The number of copies of a string of nucleotides, their position in the chromosome and their function, if any, determine the effect on the person—from inconsequential to fatal. A duplicated segment may lie adjacent to its founder segment or it may have been repositioned elsewhere, even inserted into another chromosome. Although the studies have only begun, copy number variation is expected to account for numerous good and bad differences in the health status of individuals.

A positive case is the increased number of amylase genes that raises the concentration of human salivary amylase helping to break down starch in foods as we chew. Chimpanzees have only two copies of the amylase gene (*AMY1*) which are sufficient for their low starch diet of fruit, leaves, and insects/meat. In contrast, some human tribes have more than a dozen copies providing a corresponding increase in salivary amylase and better digestion of starch. Remarkably, there are fewer amylase genes in some African tribes that herd livestock (they have low-starch diets of milk, blood, and meat) compared with African tribes that mainly eat starchy roots and tubers. The Yakut tribe of the Arctic whose diet is primarily fish, also have fewer copies of the amylase gene. Evidently having more copies of the amylase gene provides a positive selective advantage for those who subsist on starchy foods (Perry et al., 2007).

Ignoring known and unknown differences in copy number, compared with two humans there is a minimum of 10 times as many nucleotide differences between human and chimpanzee DNA—36 million bases. This figure may provide some relief to those initially offended by the discovery that the bases in human DNA and chimpanzee DNA are 99% the same. It turns out that human and mouse DNA are also quite similar. Ninety-nine percent of mouse genes have human counterparts. The general rule is that genes are remarkably similar within an animal group like mammals. Let's look at an example of one gene that is highly conserved across animal groups.

Without exception every normal human has the same identical set of 716 amino acids that make up the FOXP2 protein. Expressed at several sites in the brain, the *Foxp2* gene is necessary for the proper articulation of human speech, as revealed by the speech problems of one rare family in which some unfortunate individuals have the amino acid histidine substituted for arginine at position 553 due to a single nucleotide substitution of A for G (Lai et al., 2001). Considering its involvement in language articulation and comprehension it may surprise you that the *Foxp2* gene is not unique to humans. FOXP2 protein is almost identical in other primates, in rodents, and even in song-birds. Even though the human and chimpanzee lineages diverged some 5-6 million years ago, their FOXP2 proteins differ by only two amino acids out of 716. These two amino acids and just one other are the three differences in the FOXP2 protein of humans and mice even though they diverged more 70 million years ago. FOXP2 differs by eight amino acids between humans and zebra finch song birds whose common ancestor lived more than 300 million years ago. The facts that the FOXP2 protein has changed little in millions of years and is wholly uniform among normal people suggest that it has key roles. FOXP2 probably influences the readout of several unidentified genes in brain cells. Infant mice with one disrupted copy of the *Foxp2* gene fail to make the normal ultrasonic vocalizations in response to separation from their mother (Shu et al, 2005). Mammalian *Foxp2* seems to be one of several genes that contribute to auditory social communication.

Are genes masters of all? By dictating both structure and function, genes are largely responsible for each organism's appearance and behavior. Natural selection acts to favor those clusters or ensembles of parental genes which improve the vitality of the offspring. "Bad genes" lower their owner's viability. The success of "good genes" is evident in the high rates of survival and reproduction as the offspring develop. Successful organisms express successful genes.

All of this seems harmless enough for the individual, except that genes may enhance their replication across generations at the expense of the individual organism. Most dramatically in some species like spawning Pacific salmon or a male praying mantis the sex act is fatal for the individual but highly successful in passing on the genes to the next generation. These one-and-done affairs are victories only for the surviving genes. Accordingly, we need to view genes as successful replicators that construct a vast array of temporary vehicles which house and protect them. Those vehicles are the bodies of organisms like single microbial cells, plants, animals, and even humans. Human bodies protect and sustain genes generation after generation. We are but temporary repositories of genes which are repeatedly duplicated and passed on from adult to newborn babe. Only humans are potentially smart enough to counter genetic programs that control and use us as disposable sexual vehicles—temporary containers carrying genes into the future. You are in the clutches of your genes.

Kinships in life. Historically, humans share deep kinship with all animals because we have emerged from common stock, from the same roots. Most of our genes have stunningly ancient origins. That is the central message of molecular studies of evolution—animals have a shared history indelibly recorded in their DNA for all to see. A small change in turning genes on or off during development can result in major changes in an organism's form and function. Many of the giant strides of evolution are the sum of numerous small steps over many, many years. New animal forms often arose as a result of small changes either in existing genes or in the sequences of DNA that regulate gene expression. Species differ not simply in their catalog of genes but also in when and where particular gene protein products are generated and how they are subsequently used. In this common evolutionary motif the same protein is deployed in different ways in related species—the same chemical has varied applications.

In their reverence for life, the Buddhists got it right. When you kill any creature, including insects, you are destroying thousands of genes that closely resemble your own. It doesn't matter whether it is a lion or a mouse, a clam or a louse, a dove or a hawk, a fly or a flea; even brewer's yeast or a bacterial nitrogen-fixing microbial beast has many genes like yours. Because the lives of organisms are all based on the three-letter DNA code for amino acids, Darwin seems to have been on the right track in thinking that life on Earth arose from one primitive form, a singular ancient source. There is an unsettling insecurity in the probability that all life currently surviving on Earth had a single origin billions of years ago. Looking back on the origins of organismal diversity, geneticists credit sudden duplications of long strands of DNA and the gradual selection of small mutational changes over vast amounts of time for our Earth being alive with "endless forms most beautiful and most wonderful" (Darwin, 1859). A genetic universality lies hidden beneath the many splendors of life on Earth.

4

&

Human Needs Construct Religious Beliefs

The Psychological Roots of Religion

The hallmark of religion is belief in supernatural features or forces. Humans tend to be drawn to the power implicit in religion as a way of coping with the challenge and uncertainty of life. We may embark upon a religious quest to root our lives and our relationship to a supernatural power that might lift us to everlasting repose in Heaven. At its best, religion encourages us to contemplate life on Earth fully, as we search for meaning. Every institutional religion is a human creation whose doctrines and rituals target human problems. The doctrines of the most popular western religions assert an interventionist God surrounded by narratives, rituals, sympathy, and faith. As part of religious faith one may envision a pleasant future, now and forever more. Before considering some positive features that provide vital sparks for religiousness today, let's first examine so-called existential problems and their primal anxieties which have spurred religious quests since ancient times. Historically, religions address concerns about: how to cope with suffering and evil on Earth, the torments of our egos, and our hope for an afterlife to overcome death.

Evil and suffering. Whenever life is harsh and unfair it may motivate one to seek relief in a religion. "Suffering" can only occur in humans and other animals with brains capable of feeling or experiencing pain and hurt. It is natural to get upset at physical ailments that cause suffering. Why can't medicine protect us from nasty microbes and genetic diseases, chronic pain, depression, migraine headaches, and the like? Accidents and the crimes of evil people come without warning or justification. (I consider "evil" to mean harm perpetrated by malicious humans.) Why does it seem that so many good people are victims, while the mean spirited and guilty often go unpunished? Religion helps us cope with life's injustices and physical suffering. Chapter 8 provides a thorough discussion of evil and suffering and how they have long threatened to contaminate the purity of God.

Egos in jeopardy. Battered egos are another major source of distress that religions seek to ease. Our ego represents our sense of status and how we value ourselves in relation to others. Big egos sport that "me first" attitude. Envy, anger, pride, greed, personal slights and mockery, opportunities unfairly denied, failures in romance, gnawing insecurity, and rootless boredom; all of these may torment our egos. Rather than strutting about with unbounded confidence, it may be wiser to feel some twinges of unease or a lack of fulfillment on occasion. However, if our ego remains deflated, we may turn to religion to seek relief from feelings of inadequacy or emptiness. Should the ego deflate completely we may sink into lethargy, accompanied by a lack of purpose and a sense of failure, a feeling of disembodied anxiety and deep blues—the collective hallmarks of depression. We can scarcely feel "pumped," since there is no air in our balloon. In ironic contrast, living with inflated self-concern can mean a life of perpetually unsatisfied needs—we can become insufferable egoists, always craving more. No matter how they arise, tormented egos are the source of altogether real psychic pain. All manner and matters of tormented egos fill the counseling hours of the clergy. Wise clergy can offer helpful advice.

Death. In all cultures death is an undeniable concern. In his observations of indigenous Mayans, Bishop de Landa noted in 1572 that, "These people had a great, excessive fear of death and demonstrated this in all the many things they did for their gods, which were for nothing else but to be rewarded with health, life, and food." Even today none of us is indifferent to our own mortality, to the inevitability of death. Of course, for the most part to be young is to feel indestructible. Full of zest and vigor, why should youth anguish over something as remote as death? Isn't that more of a concern for the elderly and the ill? Few teenagers dwell on death, unless it has tracked them down and cruelly swept away a close friend or relative, or depression has managed to darken some days with personal foreboding. Paradoxically, images of death's mysterious obliteration of life create a surge of excitement that makes depictions of violent death an astounding commercial success in movies and TV.

In relentlessly trolling for larger audiences, the media are in no mood to shield us from death. As noted by the former editor of the New York Post, Pete Hamill, "You don't pick up a tabloid to find out about the sisal crop in Malaysia. The best story of any type in a tabloid is murder at a good address." Every day crass producers find another hapless fatality as the lead story for TV and newspapers—with a preference for a victim near your neighborhood so you will say, "That could have been me." On slow days, when your local scene is peaceful, count on the mass media to thrust tragedy into the living room by importing death and mayhem from some remote piece of the world. As Aldous Huxley, the author of *Brave New World*, quipped about the coverage in the morning newspaper 50 years ago, "Has anything more than unusually disastrous happened overnight?" The media hopes tragedy will give us an exciting chill, as a rush and then relief that "Thank God it was someone else who got nailed." Their warning "these graphic images may be disturbing," is often a ploy to keep you looking. While we are free to ignore the daily catastrophes that pass for news or the mayhem of Hollywood movies, we often absorb them because death is gripping.

To counter the anguish over the inevitable death of our flesh, religions offer up hope for a pleasant afterlife in Heaven or alternatively a first-rate reincarnation. We are given three basic after-death options. We are either headed for Heaven or for Hell—or equivalently, for a good or a bad reincarnation. Or death may bring eternal repose—each of us is a fallen dewdrop returning to the stream of life. Sooner or later, the meaning we give our lives will be tinted by how we view death, or more consequentially, our fate after death. Suppose you knew how and when you were going to die. This knowledge would allow you to map out your plans for life more sensibly: when to have children and how to provide for them, when to vacation, how soon to retire—planning all of the big stages of your life. Yet, is it a favor to be told when you are going to die? Imagine what you would do. Wouldn't you lie awake each night hatching scheme after scheme to cheat an impending death? Nor would it be much fun to receive advance word on your kind of fatal disease. Take Huntington's disease. This is an inherited neurological disorder of brain regions that control how we move. Gummed up with excess protein, nerve cells begin dying in earnest by middle age. At first there are months of sporadic twitching. Then, relentlessly, as more motor nerve cells die, jerking and flailing become more extreme and increasingly unpleasant. One's thoughts and personality also deteriorate. Death is inevitable in five to ten years. It is no wonder that one of the pioneering investigators of Huntington's disease has had no interest in being tested to determine whether she will suffer this ghastly fate. Understandably, she has avoided taking the simple test to show whether she herself bears this dominant mutant gene that runs in her own family. (She had no plans for children.)

As a cousin of death, sleeping is a simple test to see whether you are seriously anxious about what lies beyond death. Assume that you have no close friends or relatives, so you need only think of yourself. Would you care as you fall asleep that you might not wake up? The following morning you will either be alert for the day's adventure or you will have no awareness whatsoever. If going permanently to sleep in this manner doesn't particularly concern you, you may not be driven to religion by an obsession about life after death.

It is evident that residing within each of us is the urge to ward off the major negatives of life: evil harm and suffering, the torments of the ego, and the specter of death. Institutional religions relieve existential stresses with elaborate, sometimes flamboyant, efforts to dispel our fears and guide us along a path of fulfillment. We are often moved to act against deep threats, rather than enduring them in resigned silence. As individuals we actively try to protect ourselves from harm by controlling the immediate circumstances around us. And because humans can anticipate the future, we also find ourselves planning ahead to avoid looming threats, including planning for such elemental needs as thirst and hunger. An understanding of how things work provides us with considerable protective power because once informed, we may frame sensible predictions that allow us to sidestep future harm. Most humans have a need to understand and control their environment, to live communally and attach meaning to their lives (Spilka et al., 2003). While none of these factors insures religiousness all of them can feed into it. As we will discuss, the more positive incentives for religious observance are likely to be especially important in developed countries because their higher standard of living, healthier lives, and reduced superstitions have deflated many of the ancient negatives.

Remembrance: From Childhood to Adulthood

The roots of religious belief are nurtured by childhood fairytales, stories of miracles, and fantasies that precede mature ruminations about the meaning and purpose of life. As children we may have had no greater delight than to escape into a world of fantasy. Peter Pan, pixies, magic dragons, tiny talking animals, angels and fairies, spooks and goblins, Santa Claus, Pipi Longstocking, Alice in Wonderland, the Easter Bunny, Sesame Street's Cookie Monster—every healthy childhood has such delights and fantasies. Many children invent their personal secret friend, or secret place—me and my shadow, or my secret garden. Even as adults, we find fantasies intruding into our thoughts wherever youthful spirits persist as part of our makeup. The angelic and saintly supernatural beings that populate many adult religious beliefs help to link the world of adult concerns with the comforting myths of childhood. For many who have happy childhood memories from Sunday school and church services, the religious activities of today readily reawaken those secure childhood feelings.

Thwarting Anxiety with Religious Reassurance

America immediately after World War II. Several lingering concerns carried over into the American social climate of the late 1940s. Residual malaise from the deprivations of the Great Depression of the 1930s and the trauma of World War II (1940-1945) fused with growing fears of communism and the prospect of atomic warfare. In an increasingly mobile society, employment opportunities took American families away from their familiar small town settings. As Americans moved to new surroundings in burgeoning suburbia, they simultaneously sought the comfort of familiar religious denominations.

Churches and their familiar liturgy provided safe harbors during the transition of moving. Displaced parents found spiritual moorings for their children. Beyond the anxious motivations of the uprooted, several positive factors also impelled religious observance. In addition to recapturing some of the positive feelings of a religious childhood, church attendance and active participation in religious observances and church social activities knitted families together.

As a refuge for worship, a looming church or cathedral encouraged reassuring expectations in troubled times. Churches raised the odds of forming friendships based on shared values. By 1950 church membership had soared more than 50% in the United States. For the first time "IN GOD WE TRUST" was stamped onto

American money and "Under God" inserted into the Pledge of Allegiance. The ruddy embrace of church fellowship was a welcome antidote to the pallor of loneliness.

Continuing returns from religious experience. Contemplation and prayer, giving thanks, receiving sympathy and condolences, and grieving in a serene religious setting are spiritual activities that encourage personal transformations and restore one's outlook and vitality. If we feel we have lost our way, we can be restored by the rhythm of religious services that invoke scriptural reassurances anchored to sacred creeds.

The desire to be happy is a positive incentive for adopting a religious lifestyle, assuming that the intention goes beyond the mere pursuit of self-indulgence. A life that quests for fulfillment through personal achievements such as raising admirable children and having loving relationships in a religious community is a blend of several positive motives for spirituality. For some, direct experiences with the harmony and beauty and serenity of nature prompt spirituality. We may respond to the raucous vitality of a kingfisher's call as it swoops along a riverbank or the repetitive hoots of the yellow-billed cuckoo on a lazy summer day. That these same species are on the dinner menu of the duck hawk (peregrine falcon) provides the spiritual lesson that nature's creatures are ephemeral.

If you have a positive outlook, you may find that religion flows out of personal success in life. You may have been blessed with a loving family, a successful career, good health, and several fortunate escapes from potential disaster. As a New York Yankees baseball star famously said, "...I consider myself the luckiest man on the face of the earth," when forced to retire on July 4, 1939 with ALS (amyotrophic lateral sclerosis, soon to be named Lou Gehrig's disease in his honor). Count your blessings—most Americans have much to be thankful for—as the eye-opening privation in developing countries can reveal. However, by itself such satisfaction-based spirituality may not prompt church attendance. It is in seeking relief from difficulties that many individuals find their way into the embrace of a consoling church.

Box 4.1 Sectarian Fragmentation and Creation Stories Are Inherent in Religion.

Doctrinal squabbles may cause religious denominations to break apart into new sects. The Mennonite community, which settled in Belize only within the past century, has already split into multiple groups over seemingly inconsequential matters of conformity. One suspects the desire for power and control contributes to such splitting. Elsewhere in this region, the earlier fragmentation of Mayan communities into isolated tribes is more complete because it began many centuries ago. Divergence in neighboring Guatemala alone has led to 22 Mayan tribes, each with their own colorful costumes and a language unintelligible to any of the other tribes. The largest ethnic group is the Quiché of about 1 million people today. We would know much more about Mayan cultures had the conquering Spaniards not burned all but four of the "hieroglyphic" books which recorded their cultural features. In the words of Fray Diego de Landa (1572, p. 136), "We found a great number of books written in these letters of theirs and, because they contained nothing in which there was not superstition and falsehoods of the devil, we burned all of them. The Indians felt this very deeply and it gave them great anguish." Fortunately for historians, in the mid-1500s three unknown Mayan authors pooled their knowledge of oral traditions to write down mythic Mayan creation stories in the Popol Vuh. They wrote in Quiché Mayan transliterated into the Spanish alphabet which the Spanish priests had taught them. A century and a half later in 1703, Friar Francisco Ximénez sustained this thin thread by making a Spanish translation of the last remaining copy of the Popol Vuh. Its sublime text relates "the story of the emergence of light in the darkness, from primordial glimmers to brilliant dawns, and from rainstorms as black as night to days so clear the very ends of the earth can be seen" (Tedlock, 1996, p. 15). Interestingly, the Mayans authors of the Popul Vuh believed that by accounting for everything, their creation story, their "prior word"...took precedence over "the preaching of [the Christian] God."

Features Shared by Most Religions

Religion can provide a fabric of norms for acceptable behavior and foster loyalty to the clan under God's protective shield. Every religion offers a creation story to explain the universe and life. Most religions offer a rich liturgy, often beautifully musical, coupled to the hope of personal rapport with the supernatural, notably through answered prayers. To sustain beliefs and relieve humans of the torments of evil, physical and emotional suffering, and mortality, all religions offer six crucial elements:

1. Empowered supernatural forces, even a personal God.
2. Narratives, including a creation story that relates human lives to the supernatural. (The successive spiritual reincarnations in eastern religions can be seen as an alternative to a master plan for Salvation in traditional Christianity.)
3. A set of religious practices that include prayers, rituals, celebrations, self-denial, and sacrifices.
4. Doctrines and beliefs that include a moral code of best practices and condemned practices.
5. Fellowship of the committed. Social interactions among members can strengthen individual belief and determination through various mutually reinforcing commitments cemented by emotional experiences.
6. Material symbolism frequently includes crosses, stars, statuary, regal dress, or reverential paintings, and an inspirational site or place of worship that can be uplifting, humbling, and intimidating all at once. Religions have histories that cycle between smashing idols and venerating them.

Religious participation strengthens: i) self-discipline, ii) social bonds, iii) belief in a common heritage, and iv) a feeling of well-being. These general features have survival value. Anthropologists emphasize that religious beliefs are often associated with social structures such as whether relatives live in clusters, the likelihood of attack from neighbors, whether food is obtained by fishing, hunting, gathering, or agriculture, the tribe's social stratification, and the split between private and communal possessions (Swanson, 1960). Various studies indicate that religions are able to confer health benefits, such as reductions in the use of nicotine, alcohol, and other addictive drugs. This is not to argue that a healthy outlook is only available to those in "approved religions." *Some secular considerations.* Non-religious people with drug-free life styles and confidence in their world-views are also shown to be healthier (Spilka et al., 2003). Nor are wonder, fear, trust, and commitment unique to religious social interactions. Churches that support humanists can extend their moral reach. If the moral training of a religious majority results in widespread compliance with moral codes and laws, society's non-religious secular elements stand to benefit. While a Christian may firmly believe that humanists face an eternity of regret, it is un-Christian and strategically misguided to shun humanists prepared to learn from the religious. A pragmatic truce between secular and religious elements can be mutually supportive, and lead to cooperative interactions. Religious-secular engagements in a pluralistic society can reduce the arrogant excesses of each group. Certainly, both the religious and the non-religious have reason to be wary of the motives of politicians who attempt to construct public morals, including those pertaining to children. Freedom of religious practice requires that citizens should also be leery of judges, elected officials, and bureaucrats who would permit today's majority Christian religion or denomination to impose its views upon society as a whole.

The psychiatrist Sigmund Freud saw religions as escapes from reality, with the various religious institutions catering to people smitten by elaborate fables about the supernatural. (Ironically, it is evident now that much of psychoanalysis was Freud's own elaborate fable about sex viewed through the distorting lens of Victorian times.) Rationalists, like Freud, who rail against the rigidity and implausibility of religious beliefs, risk overlooking the social benefits of religion, including a moral code, extensive social support for those in need, and the reduction in stress that peace of mind can provide. The positive contributions of institutional religion to culture and community life include patronage of art and music, monuments, painting, statuary, temples, churches and cathedrals, education, health care delivery, and public charity (see Chapter 17). Nonetheless, the virtuous actions of organized religion must be balanced against the tendency to subordinate human health and welfare both to the "higher" cause of Salvation and to less lofty goals of denominational dominance. To arrive at the net result requires a long discussion, necessarily weighted by the things that matter most to you.

5

&

Western Religion Lightly: Christian Roots

Doctrinal Roots and Shoots

While it is easy to identify numerous differences among Muslim, Jewish, and Christian faiths, one should also appreciate the remarkable parallels in their scriptural narratives and their overlapping Middle Eastern origins. The Qur'anic narratives contain some of the same highlights found in the Christian Bible and Hebrew Bible. For example the Qur'an describes the first man, Adam, as banished from Paradise for eating forbidden fruit. Noah's ark saves the lucky few from a catastrophic flood. Abraham (Ibrahim) is willing to sacrifice his first son to prove his loyalty to God. Moses leads his people out of oppression in Egypt. The crucified Messiah, Jesus, ascends to Heaven. The scriptural narratives of all three major Western religions claim the prophet Abraham as their distant father. Jesus is said to have descended from Abraham through 42 generations (*Matthew* 1:17). One could imagine that this shared mythology and the deep ancestral root of Abraham could inspire an ecumenical movement to knit together Muslims, Christians, and Jews. Yet today, profound competitive insecurities spawn rancor and hostility rather than the cooperation that might have emerged from common narratives. Close genetic ancestry can easily increase rather than decrease hostilities. Since peace today remains an elusive goal, humans would benefit from a reawakening of Mohammed's remarkable gift for unifying diverse tribes, especially if it could be accomplished with fewer fights.

In about 610 CE the angel Gabriel visited Mohammed. This was the same angel who some six centuries earlier had announced to Mary that she was pregnant with Jesus. Gabriel spoke verities to Mohammed and commanded him to repeat those holy words in uniting his people. Perhaps because Mohammed could neither read nor write, his holy words were not immediately written down. It was not until several decades after Mohammed's death in 632 CE that oral narratives were written down in the Qur'an. Nonetheless, every passage in the Qur'an is considered to be the perfect, timeless, and unchanging Word of God. As much or more than fundamentalist Christianity, Islam has a tradition of spiritedly defending its Holy Scripture against skeptical questioning or even careful scrutiny by historians. It is insufferably painful to have non-Muslims pick away at the meaning of Qur'anic verses; even translations are forbidden. Perceived as an insult and defilement, literary criticism of the Qur'an strikes at the heart of Muslim respect and pride. The Qur'an proper contains 114 sections called suras arranged in order of increasing length rather than by content. Although portions are difficult to understand, the lyric beauty of the original Arabic is extraordinary.

The Triumph of the Resurrection

Christianity formed in the crucible of tribal conflicts among Jews and in the tensions between Romans and Jews. The divinity of a resurrected Christ is the foundation stone of traditional Christianity. Here I take the view that two millennia ago (2,000 years ago), as one of many itinerant preachers, it was Jesus and his followers who eventually gained crucial traction for a message that swirled out beyond the dusty, squabbling tribes in the Holy Land where Jesus lived and died. Christians viscerally resent the persistent trickle of belief that Jesus was a fictional creation—an amalgam of multiple prophets fused to appeal to our needs, analogous to portraits of God intended to swell our egos by painting "him" to look like an admirable man.

Box 5.1 Some Highlights of Early Christian History (after Baigent, 2006).

4-6 BCE	Birth of Jesus (*Matthew* 2:1; *Luke* 2:1-7)
27-28 CE	Jesus is baptized by John the Baptist (*Luke* 3:1-23)
30-36 CE	Jesus is crucified

50-52 CE	Paul's first letter to the Thessalonians
65 CE	Paul is executed
115 CE	Bishop Ignatius quotes from several early letters of Paul
125 CE	Date of the oldest fragments of the New Testament (*John* 18:31-33), now held by the John Rylands Library in Manchester, UK
135 CE	Victorious over the Jews, Emperor Hadrian renames Judaea, Palestina.
180 CE	Vehemently excluding Gnostic writings, Bishop Irenaeus starts to frame the 'authentic' contents of the New Testament by embracing the four canonical Gospels.
195 CE	Bishop Clement agrees with the censorship of Mark's Gospel and suppression of the Gnostics.
200 CE	Oldest complete copy of any Gospel (*John*)
325 CE	The Council of Nicaea votes 217 to 3 to make Christianity the official doctrine of the Roman Empire. The three dissidents are exiled.
367 CE	Bishop Athanasius attempts to destroy all "non-canonical" holy books in Egypt.
390 CE	Pagan Gauls destroyed much of Rome.
393-7 CE	The Councils of Hippo and Carthage finally agree on the contents of the New Testament, more than three centuries after the events it describes.
410 CE	The Visigoths complete the destruction of Rome.
440 CE	Pope Leo I claims the primacy of the pope as representing the "mystical embodiment" of Peter.

As they vied for converts, pre-biblical preachers fervently sought to gain influence and credibility through prophesy, (predicting the future), touting an afterlife, and performing miracles like healing the sick. (A miracle is defined as "any supernatural alteration of nature.") Jesus went beyond these popular strategies to attract adherents by periodically needling mainstream Judaism for its deficiencies. Naturally, such pointed critiques led to blustering threats from the Pharisees and other local Jewish powers, backed up by concerned Roman officials, ever nervous at possible unrest and rebellion. Contrary to the Gospel of Matthew, Jesus supported the Pharisees in their emphasis upon charity and love, but his potential for arousing the Romans must also have worried the Pharisees in power. Jewish unrest continued even after Christ's crucifixion.

Jewish dissenters were eventually intimidated by the Romans' appalling destruction of Jerusalem in 70 CE. This suppression also meant that the compilers of the New Testament would be even more reluctant to single out the Romans for the death of Jesus. The alternative of blaming Jewish authorities and the Jewish populace was a safer way to advance Christianity. But regardless of who was to blame for Jesus' death, it is far more important to appreciate how essential it was that Christ be crucified. Consider for a moment whether you think that Christianity would have taken hold to be with us today if Jesus had continued to live on, preached for some years and then, as he grew older, quietly retired to village life without crucifixion, resurrection, or ascension into Heaven. The crucifixion saga is much more than a mere flickering in Christianity's search for luminous standing. The resurrection and ascension are some of the most revered and essential beacons that illuminate the path to Christian belief. Without evidence for Jesus' sacrifice and divinity there would be no Christian Salvation—and almost surely no Christianity at all. All Christians should feel that the crucifixion of Jesus warrants a solemn hymn of gratitude, regardless of the unsavory motives of the responsible individuals. (If you are inclined to blame Judas, hold your fire until you read in Chapter 6 about his recent and remarkable makeover.) Jesus died on the cross that we might be saved. Bless this wicked iniquity.

A forlorn Jesus Christ nailed to the cross is the saddest and most prominent image in Christian theology. The early bishops elevated this iconic symbol of death and guilt, rather than offer a resurrection symbol that would have shown our hopes in a Jesus triumphantly revived, radiantly ascending to Heaven with arms outstretched not by nails but in a loving embrace of all humans. Why? Why choose the dreaded Roman cross as the symbol of Christianity? Perhaps it symbolized a sacrifice that atoned for human sins? Certainly it was well-grounded in Hebrew practice that one could atone for a sin by slitting the throat of a sacrificial animal. A pigeon was good enough for minor sins, and a lamb for major sins, but the sins of a nation required the death of an ox. As Hebrews, Paul (a.k.a. Saul) and John (3:16) famously said that Jesus had been sacrificed to atone for human sins. An outsider might find this strange since sacrificial offerings have typically been made by humans to please God. What is the meaning of Christianity's stunning reversal of roles—God sacrificing his Son for humans? Does it resemble the loving sacrifice of a person willing to stand in for a convicted relative and bear the court's legal sentence? By bearing the cost of humanity's transgressions, Jesus' sacrificial death paid the ultimate price for all of our sins. The resulting human gratitude and guilt have helped to energize and vault Christianity over other religions. Ironically, the many later Christian martyrs met their horrific fates with little assurance that in a matter of days they would be revived like Jesus and ascend to Heaven for all eternity. The fateful theological choice to emphasize torture and death, and sin and guilt, would fracture Christianity forever.

After the grisly crucifixion of Jesus, most of his apostles returned to their previous occupations, instead of traveling about proclaiming the miracle of his ascension and the truth of Christianity. Over the years some active disciples, notably Paul, in speaking and traveling widely, generated narratives so nourishing to the hopes of Gentiles that Christian roots began to grow and take hold. This process took some decades. The interval between the death of Jesus and the beginning of Paul's missionary career is a murky period of Christian history (30-50 CE). Since only 10% of the population could read and write, it is likely the legacy of Jesus relied on oral traditions that carried forward the gist of his teachings from an earlier time. Certainly if "Christian" biblical documents were written before 50 CE, they have not been discovered.

It is thought that soon after Jesus' death his followers retreated to their own separate camps. At times the disciples and relatives of Jesus squabbled in their competing attempts to roust up followers for what were initially rather traditional Jewish views. Had the views of any of these Jewish groups deviated substantially from Judaism of the time they would have risked losing followers. As the half-brother of Jesus, James was one of the more important figures in sustaining early Christianity (Tabor, 2006). The Jerusalem Christians attached to James were ordinary Hebrews who expected to see God reign at any moment. Having similar hopes, the vegetarian Ebionites followed the Torah exclusively. As one of several rival groups, the Ebionites esteemed Peter and abhorred Paul. John's group separated dramatically from the Jews in declaring that Jesus and God were one. Although Matthew's Christians maintained traditional Jewish practices, they steadfastly believed that Jesus was the Son of God by a virgin birth. Paul's followers viewed Jesus as a noble human less divine than God. Most importantly, to separate Christianity's doctrinal path from Jewish laws, Paul argued that Jesus already existed in some manner before the Torah's laws were adopted. A fateful meeting of John, James, Peter, and Paul in Jerusalem crystallized two visions of Christianity. It was agreed that Paul would evangelize the Gentiles offering Christianity as the "new covenant" promised by the prophet Jeremiah (*Jer.* 31: 31-34) while the other three would engage the Jews and sustain Jewish traditions (*Gal.* 2: 11-21). In time Gentile converts came to substantially outnumber Jewish converts (Johnson, 1976; Rigas, 2004). Paul's vision for Christianity would triumph.

The overarching impression of this tumultuous period is that practicality triumphed over divine revelations in constructing Christian belief. Immediately after the death of Jesus there were in Jerusalem as few as 100 Christians fragmented into tiny competing sects. Success depended upon jelling their beliefs into inspiration and spirited acceptance. Paul's foundational beliefs depended upon the death, resurrection, and ascension of Jesus. These potent selling points were obviously unavailable to Jesus while he was alive and preaching. In brief, Paul asserted that the resurrection of Jesus meant that other humans could follow in his wake, provided they were redeemed in Jesus through baptism, Eucharist, and faith in God's grace. In the end, the belief structure of Christianity was not so much provided by Jesus, who never professed a desire to found a religion, but by the efforts of Paul and the theology set forth in his letters and related writings. Indeed, Paul and his group of disciples wrote roughly half of the New Testament.

Over several generations various narratives were edited and melded together, extensively revised, and further embellished to create stunning written accounts of the life of Jesus and his teachings. While Roman officials viewed Jesus as a nettlesome preacher, to his most adoring followers he was the Son of God—a divinity who was crucified to free us from our sins. Compellingly revived from death, Jesus ascended into Heaven, having already promised to return shortly to gather up the deserving faithful and shepherd them to Heaven. This glorious account of Jesus triumphed over the teachings of other preachers largely because the saga of his resurrection proved his divinity to his acolytes. To see and hear and touch a revived Jesus would surely have been awe-inspiring. After all, you can't touch and feel the flesh of a ghost. Imagine those who directly and intimately confirmed Jesus' substance. What an inspiring moment in Christian history (Johnson, 1976)!

As a notable contemporary preacher, John the Baptist (about 5 BCE-32 CE) zealously baptized a large following, including Jesus, in order to wash away the sins of man. (As the Son of God or God transformed, it seems unlikely that Jesus had any sins, so his acceptance of baptism may have been a gesture of solidarity and nascent leadership.) How was it that the influence of the popular John the Baptist was eclipsed by the growth of followers of Jesus? In part Christianity benefited mightily from the tireless proselytizing by Paul and other dedicated acolytes as they gathered followers and built churches. Moreover, while the legend of Jesus' resurrection was able to spread persuasively, it was rather difficult for John the Baptist to appear divinely resurrected after Herod had his head plopped onto a platter—a gruesome image that became the centerpiece of numerous tableaus in Christendom.

Buffeted by stiff competition for human hearts, Christianity's fate hung in the balance for centuries—was it to survive or die out along with hundreds of other minor sects? Emperor Constantine's decree in 312 CE that Christianity was the official religion of the Holy Roman Empire provided a tremendous boost in legitimacy and power. Nonetheless, almost immediately there were spirited debates about a recurring paradox in Christian theology. If it was correct that only one God existed, how could there be two divine figures: Jesus and God? To resolve this festering concern, Constantine gathered the bishops at Nicaea in 325 CE where they generated a consensus, a forerunner of the Nicene Creed. Even while doubts disappeared about Jesus' divinity, some issues remained about the God-man fusion. Clearly, it would not do to continue in the Greco-Roman tradition where the gods were often half-man and half-God. This model was also too close to bestial chimeras. Further, a God-human chimera would raise questions about which parts died on the cross. Eventually, some 300 bishops agreed that Jesus was fully God and fully a sinless human able to have experienced suffering like an ordinary mortal. This doctrine of Incarnation is one of the distinctive cornerstones of Christianity. Jesus is the bridge from humans to God—from the natural to the supernatural.

With the Nicaea accord only briefly mentioning the Holy Spirit, a divine emanation from God that could enter humans, it remained for Augustine (354-430 CE) to mold the concept of the Trinity for Catholicism. Today the concept of Trinity rides uneasily on the backs of Christian denominations willing to cope with how the Father, the Son, and the Holy Spirit can co-exist within a monotheistic religion. Water is a provocative analogy for the concept of the Trinity, since it exists in three states—liquid, solid, and gas. For the water metaphor to buttress the monotheism of a trinitarian religion, God must at times be simultaneously composed of three states, solid the Father, liquid the Son, and vapor the Holy Spirit—three forms of holy water. A tripartite composition helps to dodge the dilemma of how a dead Jesus could revive himself.

In the concept of original sin, Augustine also offered a possible solution to the vexing problem of an all-good God allowing evil and suffering. As a sexually transmitted affliction that "accounts for evil," original sin makes its obligatory entry with the sperm during intercourse. He viewed the sexual act as an indulgence so pleasurable that God was sadly forgotten during that exalted moment. Yet, in praying for his own celibacy, Augustine wryly remarked, "Lord, please give me chastity—only not yet" (Chadwick, 1991).

Box 5.2 The Real Jesus?

Institutional Christianity has always been concerned that some damaging evidence, like an ossuary containing the bones of Jesus, might turn up to undercut the standing of Jesus. Many sources, including the Qur'an, take the view that Jesus was an admirable, yet non-divine, historical figure. Had Jesus actually survived his crucifixion, or had he died without being resurrected, it would eliminate the best argument for his divinity. And if Jesus had children through a wife (Mary Magdalene?), it would undermine Jesus as a model for the celibate priesthood and popes as non-hereditary leaders. Although marriage was the rule for Jewish rabbis of the period, there is no definitive evidence that Jesus was married—but the rumors still swirl. Baigent (2006) has recently argued that a married, non-divine Jesus survived a botched crucifixion. Baigent saw and held two letters that he believes Jesus wrote to the Jerusalem Supreme Court (Sandedrin). His story begins in 1961 when an acquaintance found two Aramaic documents long buried in the cellar of his house near the old temple in Jerusalem. Dated with puzzling precision to 34 CE by the context of mysteriously unspecified objects nearby, these two letters were from the "Messiah of the Children of Israel." Charged before the judicial authorities with calling himself the "Son of God," the letter writer denied in court that he was divine. Was the writer Jesus, and was his denial given under duress? It is not unusual for similar tantalizing, incompletely analyzed documents to surface and then disappear from view, as these did, either because their owners wanted more money or because shadowy figures have suppressed or destroyed the material in order to protect religious dogma, or because the entire affair is a scam. You can expect rumors to persist that a mortal Jesus narrowly survived crucifixion to retire in a place like Egypt with his wife and children.

Precautions and Preaching in Biblical Times

All four canonical Gospels of the Bible mention the secret practices of Jesus and other prophets. Before performing a healing, Jesus would arrange a discussion with the afflicted person either in private or together with a few disciples. And after the miraculous recovery, the healed person typically followed prior instructions to remain silent. Doubtless some of this secrecy prevented skeptics from carrying out independent interviews and examinations to confirm it as a miraculous healing of a bona fide affliction. In addition, the quite reasonable fear of retribution from the authorities also prompted secrecy. The Romans were likely to grab anyone who claimed to be the Messiah or whose influence otherwise seemed poised to stir up public unrest. The Romans were well aware from the remarks of preachers and the buzz of crowds that a Messiah was expected to free the Jews from their most recent bondage—Roman domination. Jesus and his associates rarely revealed their whereabouts between gatherings. Secrecy and precautions against a crackdown by Roman or mainstream Jewish authorities were advisable in a Holy Land much more inhospitable to free speech than it is now.

Throughout the ages a preacher and his followers have had set roles. The responsibility of the preacher was to convey the canon, the authentic set of principles, garnished with persuasive narratives. The acolytes were to accept them. And as the arbiter of the true faith, the preacher provided the definitive interpretation. Often the meaning of parables was intentionally left murky. Followers had few opportunities to probe and question the preacher, for critical thinking was considered unhelpful. Followers were also discouraged from quizzing the recipient of a healing. Preachers would make it clear that, "I am your authority, and my messages and revelations are your highest priority. Therefore, you must believe in the canon and me. Sever your connections with competing loyalties, if necessary separating from any friends, relatives, and family who are non-believers." Persuasion became a matter of having faith.

Box 5.3 Cross-cultural Evidence for Basic Human Spiritual Needs: The Remarkable Overlap of Pagan, Old Testament, and New Testament Myths.

The miracles of Jesus recounted in the New Testament were similar to those described earlier in Old Testament narratives. Biblical tales of angels and other divine persons coming to Earth and re-ascending to Heaven are reminiscent of the shuttling of the Greek gods who in one form or another frequently commuted between their celestial abodes and Earth. Many aspects of Christian beliefs were derived from pagan practices. This is why many centuries later in the new world of the Americas it was easy for Spanish conquistadors and priests to layer Catholic observances on top of the seasonal pagan rituals of the Incas, Aztecs, and Mayans. For example, baptism was a well-established practice for the Mayans in Mexico, according to the 16th century records of Bishop de Landa (1572; Pagden, 1975). All Mayans were baptized, or in their words were caputzihil, "reborn." (kaa' sihil in modern Mayan). The baptism of children was a universal cleansing that prevented sin. The priest splashed the child with holy water from a bowl and cast incense into a fire. Adolescents with sin had to confess before receiving baptism, which was intended to guarantee a good life on Earth and an afterlife with food and drink. The Mayan priest also purified the dwelling of the baptized person and drove the devil from it. Adults routinely used incense and various idols to defeat evil. Since evil and sin were believed to cause sickness and death, adults regularly confessed their sins to a Mayan priest. Confession of sin and ritual fasting were commonplace. Wouldn't you agree that the tone of such Mayan rituals has a familiar Christian ring? Comparable rituals, including baptism, also have pre-Christian roots in the Mediterranean region. In Greek mythology the infant Achilles was dipped into baptismal water. Sadly, he was held by one foot which remained dry, accounting for the later tragic injury to that heel and today's popular expression, "Achilles heel."

From Sacrifice to Schism

Pagan rituals in earlier times subjected animals and humans to sacrifices. Christians themselves endured several centuries of sacrificial persecution that both entertained the Roman populace and quieted official concerns about an upsurge of Christian unrest. In medieval times the Christian church scaled up the Roman tradition of human sacrifices. For centuries Christian clerics proudly rescued thousands of unfortunate souls from eternal damnation by killing heretics, including women and children infected by satanic spirits. Absolute belief in faith can bear bitter fruits.

In its early phase, Christianity gained strength by bringing widely scattered churches into the fold. In 1054 CE, after nearly a millennium of unity, the Christian Church suffered a momentous schism into Eastern Orthodox and Roman Catholic divisions. In proclaiming the preeminence of their roles, Eastern and Western popes exchanged letters of mutual excommunication. Centuries later, Martin Luther prompted further schisms when he posted his 95 theses on the door of All Saints' Church in Wittenberg, Germany on October 31, 1517. Luther's disgust with the concept of paying for the nullification of sins and entry into Heaven led him to declare in his theses that "An indulgence can never remit guilt; the pope himself can do no such thing." He also proclaimed that "every man is a priest" by which he meant no cleric stood between man and God. If the pope was no longer an unassailable authority and the priests were no longer unavoidable middlemen, were the clergy even needed as authorities? The Catholic Church continues to assert its priestly authority. For evangelical Christians, and especially for fundamentalists, guidance comes from biblical scripture embraced as the authoritative Word of God. For Quakers and some others, guidance can be found in one's Inner Light, in meditation and introspection. Ultimately Luther's Protestants would divide and fragment into thousands of denominations with religious practices adapting to local conditions. Modern day charismatic religious prophets, especially of evangelical, Pentecostal, and other fundamentalist persuasions continue to spawn sects and fill churches with persuasive promises and revitalized messages. All of us seek personalized reassurances, especially about our continued suffering on Earth and the prospects for an afterlife. How many Sundays have been spent sampling churches in the hope of finding a denomination and pastor closely aligned with our spiritual needs?

Box 5.4 Quakers or the Religious Society of Friends.

Quakerism began in the 1650s in Britain as an offshoot of Anglicanism. Like many religious minorities Quakers made their way to America in order to escape persecution. Regrettably, abuse by the Massachusetts Bay Colony drove the Quakers to Pennsylvania under the leadership of William Penn. In an earlier exchange with King Charles II of England, William Penn had declined to remove his hat, saying, "Friend Charles we only remove our hat for the Lord." The King replied, "Friend William, then I shall remove mine, for in this place we only allow one hat at time." This mellow tale illustrates the Quaker belief that all people are equal under God. The four cornerstones of Quaker belief are equality, peace, honesty, and simplicity. As pacifists subject to alternative service, Quakers were often assigned M*A*S*H-like duties as ambulance drivers and medics at the front lines during World Wars I and II. Their achievements were recognized with the Nobel Peace prize in 1947. Equality was central to Quakerism at its outset in the mid 1600s when women had the same privileges of speaking as men. Quakers were well-known for their honesty in business dealings, their simplicity in dress, and an unpretentious life style. God shines as an Inner Light in each person to be revealed during meditation. The traditional Quaker Meeting is an hour of totally silent meditation in an undecorated hall overseen by seated elders, but no pastor. If so moved by the Spirit a few elders might rise at the end of Meeting to profess some insight. As you can imagine, the coupling of a simple lifestyle with services that lack music, liturgy, and a pastor means that Quakerism lacks the emotional vibrancy of most modern religious services. That may help account for its small membership—fewer than one million Quakers worldwide. But because they believe their Inner Light calls them to action, Quakers have been influential as founders of important humanitarian organizations including Amnesty International, Greenpeace, and the Oxfam charity.

Until the 1960s monthly attendance at services was required of students at Haverford College, which along with Bryn Mawr and Swarthmore Colleges is a distinguished Quaker liberal arts college in suburban Philadelphia, the heart of Quakerism in America. It happened in one of the Haverford College required religious services that the diminutive Quaker poet Bacon Evans was in attendance. Dressed in the Quaker's traditional gray collarless costume with a flat-brimmed hat, he rose and said reflectively, "I note the students seem particularly restless today (they had been sitting for an hour in complete silence). This reminds me of two skeletons in the closet, when one said to the other, 'if we had any guts we'd get out of here'." Advice taken.

Christian Creeds Summarize Core Beliefs

The Apostles' Creed is an admirably compact statement of the core beliefs of traditional western Christianity. It was derived from questions and commitments that formed part of baptismal ceremonies carried out as early as 200 CE and adopted in about 390 CE. The Nicene Creed is an augmented version of the Apostles' Creed. It enjoys the most widespread use in Christendom. The Nicene Creed is chanted at Eucharist, the ritual ingestion of Christ's flesh and blood. It establishes God's omnipotence when it says: "I believe in one God...maker of Heaven and Earth, of all things visible and invisible...through whom all things came into existence." And it claims for itself the highest standing among churches as the "one holy, Catholic and Apostolic church." Christian creeds are compact statements of the importance of the divine Jesus who, as the Incarnation of God, spoke through Peter and Paul and other prophets in order to spread the word about his sacrifice that made Salvation possible.

If there are only a few standardized creeds that lay out widely accepted core beliefs, how does one explain the existence of 33,820 Christian denominations in 2000 CE (Barrett et al., 2001)? There can't be tens of thousands of different doctrinal beliefs. Some of the denominational divergence must stem from differences in cultural and ethnic background. Churches socially isolated from each other are likely to evolve divergent doctrines and practices over time, reflecting different dynamics between ministers and their flocks. Schisms and name changes within a denomination may be driven by personalities and politics. A charismatic leader, trailed by a pod of devoted followers, may march off to plant an offshoot sect. Religious denominations that have flexible and accommodating beliefs are less likely to split over issues of dogma. To avoid fractures, brittle denominations that adhere to a rigid dogma may autocratically enforce uniform behavior, in the extreme using accusations of heresy to bully dissidents. "Heresy" is aptly derived from the Greek "hairesis" meaning "choice." Certainly those who hold to "absolute religious truth" must find the prospect of a choice among alternatives as heretical. "There is only one correct religious belief, and we have it."

The Emotional Needs of Christians

St Augustine (354-430), Blaise Pascal (1623-1662), and C. S. Lewis (1898-1963) authored three of the most famous Christian apologetics, or arguments to persuade the layman on the merits of Christianity. Orthodox Christian apologetics from Saints Peter and Paul on down to C. S. Lewis in the 20th century have adhered to the following sequence:

1. God is our true end.
2. Our moral duty is to love God.
3. By sinning, humans are alienated from God.
4. Pride and self-indulgence are the causes of sin.
5. Christ will cure sin.
6. Faith will take us to Christ and God.

Ideally faith would bring delightful contentment. The foundation of Pascal's apologetic was that humans seek religion because they are unhappy (Kreeft, 1993). C. S. Lewis (1942) also felt that Christianity was the path to contentment. Admittedly, modern lives have many non-religious advantages: humanism has provided personal rights and liberties, while agricultural and economic development has freed many from struggling just to survive. Yet, such material gains provide comfort with no guarantee of spiritual contentment. In addressing the spiritual void that materialism often fails to fill, religion has shifted from an emphasis on reducing sinning to finding fulfillment. To deal with a void unfilled by his earlier atheism, C. S. Lewis sought contentment in loving obedience to God. As Lewis said, "I was not born to be free. I was born to adore and to obey." Those were his personal needs, perhaps traceable to authority figures during his childhood.

Religious belief is driven by feelings. An overriding objective of religious practice is to establish the deeply desired emotional state, a sense of transcendent satisfaction with life and the great beyond. It is commonly implied that the commandments and the Holy Scriptures are not to be critiqued as flawed, out-of-date documents, but instead are to be almost sensually absorbed for their high poetry and spiritual aroma. To create emotionally stirring celebrations of life's milestones, religious services ritualize the joy of birth, the induction of youth into the arms of the church, adolescent coming of age, marriage, children, career, and release to an afterlife through death.

Church services often follow a scripted and soothingly familiar order and cadence, down to the measured pace of lighting candles one by one. Attendees at today's religious services may seek the comfort of these familiar rituals in a brief escape from the hectic pace of change in modern life. The devout find comfort both in public emotional evocation and in private prayer to help struggle with personal problems. Public and private religious rituals involve chanting a mantra, spinning a prayer wheel, touching an icon, counting rosary beads, saying Hail Marys, kneeling and prostrating, bowing and elevating, and taking part in praying, singing, and responsive readings. These repetitive exercises are both invigorating and emotionally arousing. The good feelings range from contentment to high exhilaration. Performances of the doxology and familiar musical rituals re-evoke childhood feelings of warm and secure family occasions. By re-arousing emotions from the past, by creating soothing or uplifting emotions through repetition and music, regular religious rituals can lighten burdens. Since God is all-knowing, he must already know of your specific needs and whether they warrant his mercy—so the point of lengthy prayers by candlelight in the cathedral may not be to tell God what he already knows, but to provide the supplicant with an opportunity to unburden and establish a soothing emotional tone. When things are going badly, you pray for help, and when things are going better as at mealtimes, you pray to give thanks for God's blessings, perhaps even for unknown blessings already winging their way to you. Prayer helps to maintain a reassuring rapport with the Almighty.

In addition to the emotive power of religious ceremonies and icons and prayer, celebratory meals are also laden with emotive symbolism. There is symbolic significance in every food item at the Seder meal during Passover that commemorates the exodus of Jews from their slavery in Egypt. The salted waters are the tears of Jews in captivity. The bread is unleavened because there was no time for it to rise in the rush to escape. The parsley sprig symbolizes the renewal of plant growth each spring. Similarly, in the Christian tradition of communion (Mass or Eucharist) the ingested wine and bread are literally transformed, in the view of some, into the blood and flesh of Christ. The cross is the symbol of torture, guilt, sin, and especially redemption. Kneeling is submission to God and to the authority of the church. Incense with its complex medley of odors is used to evoke the reassuring feelings of past observances. The memory of odors is durable, specific, and emotionally evocative. Shawls to cover women's heads may be a simple strategy to reduce sexual distractions during church services so that more fitting emotions will prevail. The religious practices we once kept as children will channel our feelings while we put down roots as adults.

Having moved to a new town, you attend church, but as you leave the service, you do not feel restored and whole. In spite of the kindly minister, the spot-on message, the colorful sanctuary, the warm and comfortable pews, the tidy and well-scrubbed celebrants, something was missing. The experience failed to lift your feelings. In attending you did not feel invested in the social fabric of that church. You will not return. Religion is about feeling.

6

Biblical Transformations and Excluded Gospels

The Old Testament

Modern versions of the Christian Bible consist of about 1,000 pages of some 40 collected narratives termed the Old Testament and about 300 pages of some 20 narratives termed the New Testament. The Old Testament originated among Bronze Age tribes living thousands of years ago. It emphasizes the period from 1300 to 400 BCE. It is a chronicle of history infused with regulations and testimonials of belief. Like all religious narratives, the Old Testament has moral instructions, along with a creation story intended to clarify the mystery of human origins.

The Old Testament began to be assembled in the 6th century BCE as a diverse collection of ancient manuscripts differing in purpose and importance. Its authors sought a chronicle to provide divine guidance for human affairs. It was no simple task to assemble the divine Word of God (Yahweh) to guide the people of Israel. These labors would last for several more centuries as the Jewish rabbis debated each manuscript's specific level of sacred inspiration. The winnowing, modifying, and editing of Old Testament manuscripts continued well into the 1st century CE. The debates over the degree of divine inspiration of early Old Testament texts would later be mirrored in centuries of controversy over the divine inspiration of candidate writings for the New Testament. In the end many passages were inspired by the desire to evangelize on behalf of Christianity.

As early as 2,000 BCE, Semitic tribes of Hebrews inhabited the "fertile crescent" between the Tigris and Euphrates rivers. A Hebrew colony established later in Egypt thrived for many years, until increasing persecution led to the harrowing escape of the Jews when guided across the Red Sea by Moses. After "40 years of wandering in the wilderness" Moses settled these Hebrews in Canaan (Palestine) around 1,400 BCE. Moses was no ordinary roustabout guide. Tradition holds that in meeting with God on Mount Sinai, Moses received an extensive code of laws, including the famous Ten Commandments, and perhaps most importantly made a fateful covenant with Yahweh. In return for Yahweh's protection, Hebrews were obligated to obey more than 600 "Mosaic Laws." Disobedience of these laws could dissolve Yahweh's protection. Disobedience was frequently invoked to explain the many hostile incursions and defeats the Hebrews suffered over subsequent centuries (Armstrong, 1993).

Local judges scattered across several districts were appointed as administrators of Judah (the southern portion of Palestine), until increased raiding prompted these administrative judges to yield to the power of nation-wide kings; first Saul about 1000 BCE, then David, and later Solomon. These relatively stable times were followed by a particularly troubled period that lasted roughly from 721 BCE to the 1st century CE. Against a backdrop of cycles of drought and pestilence, the Jews endured repeated conquests, exoduses, restorations, and revolts that sequentially involved Assyrians, Egyptians, Babylonians, Syrian-instigated Maccabean wars, and eventual subjugation by the Roman government. It is hardly surprising that the Jews yearned for Yahweh to send a Messiah from the noble lineage of David to rescue them from centuries of hardships.

The widely adopted first five books of the Old Testament recount the early history. This Pentateuch, as it is called, not only includes Genesis, with creation stories revered by many, but also contains the Mosaic Laws of the Jews that are also recognized scripture for Muslims. Sadly, even as Christians, Muslims, and Jews share ethnic origins and embrace some of the same scriptural writings, they have considerable difficulty embracing one another.

Quite apart from its erroneous astronomy, geology, and biology, the human creation story in Genesis strains belief today. Why did an almighty God create creatures he couldn't control from the start? These include not only the disobedient Adam and Eve but presumably also Satan and swarms of rebellious angels. It seems that God had to banish the obstreperous Adam and Eve from the Garden of Eden just as they were finishing the forbidden apple and strolling over to eat from the tree of everlasting life. (Clerics argue that God's system of justice would crumble if disobedience were not punished. There was no better way.) Perhaps it was just as well they missed out on this next nibble, for the immortality they and their descendants almost received would have forever packed today's already overcrowded planet with billions of immortal people. But wouldn't it be great to listen to the vast sweep of stories from really, really, old relatives?

The banishment of Adam and Eve guaranteed in the eyes of Yahweh, the Jewish God of that time, that virtually no humans who lived before Jesus were qualified for eternal life in Heaven. After all, human misbehavior was rampant. Indeed, early humans were so ill behaved that God had to make a fresh start by arranging for a huge flood to drown nearly all of them except for Noah, his four wives, and three sons. Even so, the continued sinning of humans he had created so vexed the Christian God that he eventually sent his Son, Jesus, to free humans from the bondage of their sins. As a way to abolish the virtual curse he had imposed many centuries before in the Garden of Eden, God decided to have humans crucify his Son. That is, to set things right at last, God or his Son would come to Earth in the form of a human (Jesus) and be tortured to death in order to save humans from the eternal damnation he had earlier decreed.

Christian sin is hardly exceptional. Every successful religion views sinning—lapses in obeying a covenant—as a serious matter. For those sinners who are free of any pangs of guilt, threats of Hell may lead them to return to the fold. Sinners burdened by guilt are offered more positive ways to escape from their wretched cycles of sinning and repentance. The doctrine of God's Atonement for human sin is second only to the Trinity as the most adroit and the most awkward cornerstone of Christianity. God's sacrificial Atonement on the cross is thought to liberate flawed humans from their sinful nature. Our sinful nature is somehow relieved through Jesus' suffering and sacrifice.

Aside from serious sins blocking entry to Heaven, those with non-Christian beliefs, who now make up three fourths of humanity, have no chance of being saved and may be destined for Hell, at least according to some Christians mired in rather uncharitable traditions. After all, Jesus said he was the portal of entry into Heaven. Nonetheless, even among Christian denominations the exact rules for Salvation and ascension into Heaven remain in dispute. In words that Vatican continues to employ, the 1999 document *Dominus Jesus* said that the Salvation of non-Catholics is a "gravely deficient situation." Exclusionary rules should be expected since a denomination would undercut its own standing if it agreed that one can just as easily enter Heaven through another denomination. A church's appeal and value are maximized when it asserts that its doors offer the only secure entrance to Heaven.

When examined today, the Old Testament offers some wonderfully wholesome advice laced with pitiless rules. Harsh commandments and stern advice reflected the laws of the times, notably the severe Mosaic Laws that Jesus said he intended to uphold "every jot and tittle" (*Matt.* 5:18). (If one accepts the doctrine that Jesus and God are one, a melding of Father and Son, it should hardly be surprising that Jesus would support God's Ten Commandments and other Mosaic Laws.) The more than 600 Mosaic Laws of the Jews are the laws divinely given to Moses and subsequently blended with the judicial codes of the time. Some Mosaic Laws were pretty stern stuff. "He that curseth his father or mother shall surely be put to death." "Thou shall not suffer a witch to live." "Homosexuality is an abomination." "Wizards shall be stoned to death." "Anyone who worships other gods shall be stoned to death." "A girl who is not a virgin when she marries shall be stoned at her father's door." One can claim that such unpleasant biblical quotations have been taken out of textual context but this defense loses force when similar ideas repeatedly appear in different contexts. Jesus rarely disagreed with Mosaic Law stating, "It is easier for Heaven and Earth to pass, than one tittle of the law to fail" (*Luke* 16:17). The Mosaic Laws include the Ten Commandments (different branches of Christianity recognize different sets of ten, with the New Testament only mentioning the Ten Commandments in passing). The first of the Ten Commandments says, "Thou shall have no other gods before me." This is a demand for homage to Yahweh who, the Bible says, could be jealous and vengeful. As we will see, 20th century America persecuted some of its citizens for taking the first commandment literally.

In another vestige of coarser times, the Bible provides strong support for the ritual cannibalism that is the hallmark of weekly communion. Jesus said, "Verily, verily, I say unto you, Except ye eat the flesh of the Son of man, and drink his blood, ye have no life in you" (*John* 6:53). In yet another reflection of his time, Jesus never denounced slavery; indeed, he said that misbehaving slaves might be "justifiably" whipped and "delivered to their tormentors"

(*Matthew* 18:34). "Servants, be subject to your masters with all fear" (*I Peter* 2:18). In cultural norms that date back at least 3,000 years, it was standard practice for the conquering forces to enslave the choicest of the defeated. It was not until the 18th century that western authorities got around to declaring slavery wrong.

Today with the advantage of modern hindsight one can question these coarse community standards while pointing to an inept human creation strategy and the unfairness of God's judgments as written down in ancient narratives. Genesis and the many cruel chronicles of the Old Testament reflect pre-scientific understanding in the harsh setting of the Bronze Age when mere survival was a challenge. It is a reassuring sign that today's greater level of understanding and much-improved living conditions support a higher level of compassion and decency in human affairs. In spite of the remarks of Jesus quoted above, one should not saddle him with callous enthusiasm for the cruelty in Mosaic laws. As a Jew, Jesus was in no position to defy Jewish authorities by publicly lambasting the Mosaic legal code he was taught to obey.

The Context of the New Testament

The New Testament covers the century that begins with the birth of Jesus in approximately 4-6 BCE. The absence of Christian writings before 50 CE makes it difficult to assess New Testament writings. Among the four Gospels of the New Testament, John was especially structured to lead readers to an abiding faith in Christ rather than structured to present a balanced history of events in the Holy Land. Consequently, biblical historians face the challenge of distinguishing multiple proselytizing efforts from historical accounts. Reconstructing the history of biblical events is like trying to fill in a giant jigsaw puzzle when most of the pieces of the picture are missing. Even the few ancient pieces historians possess are facsimiles—imperfect copies of unknown originals. In spite of such historical uncertainties, many individuals have claimed an intuitive sense of the intended scriptural messages.

In the gritty biblical times the hardships being endured by the tribes of Israel spawned avid audiences who clung to itinerant Jewish preachers offering hope through homilies and miracles. The Bible summarizes some oral traditions about Christ's brief life as a preacher when he managed to collect a small band of largely illiterate followers. Regrettably, even if first hand accounts of the Christian saga had been written while Jesus was alive on Earth (5 BCE-33 CE), no such writings from that period survive today. This leaves major gaps in the historical record. Almost nothing is known about the childhood pastimes of Jesus, or his family life, or even whether he went to school. Biblical historians are left with tantalizing hints, like the synoptic Gospels of Matthew, Mark, and Luke, whose substantial overlap is consistent with some kind of founding writings that historians call the Q source—a long sought account, perhaps written shortly after the passing of Jesus.

Who was Jesus? As depicted by the biblical writers, Jesus had many titles. In the Gospel of John alone Jesus had 20 titles including: the Son of man (1:51); the Son of God (3:18); Spirit (4:24); the Messiah (4:26); Christ (4:26); a Prophet (4:44); Light of the World (9:5); God (10:30); the resurrection and the life (11:25); the way the truth and the life (14:6); Prince of this world (12:31); Master (13:13); Lord (12:13); and the Holy Ghost (14:26). These honorific titles are reflections of the many roles of Jesus in the New Testament. To believers he was the Messiah (*christos,* Greek) of the Jews, redeemer of the Jews subjugated by Gentiles, ruler of a worldwide kingdom of the Jews, mystical Savior of the Jews from their sins, ruler of a Heavenly kingdom of Jewish saints, Savior of the world from sin as the sacrificed Son of God, and finally God himself who rules over an earthly kingdom and a Heavenly kingdom for the saved. These characterizations are full of divine references. The prolific use of impressive titles establishes the church's strong desire to promote Jesus' authority and divinity well beyond what Jesus himself claimed.

The New Testament as the Literal Word of God?

In continuing the tradition of the Catholic Church, Luther and Calvin considered it perfectly proper to interpret the Bible. It was not until the beginning of the 17th century that certain Protestants in northern Europe began to consider the Bible as a repository of inerrant information. Contemporary American ministers, if they lack the authoritative support of a dominant church, may understandably rely on the backing and power of the Bible as the exact and literal Word of God. Many evangelical and conservative Christians accept biblical literalism. Others realize that inerrancy implies that ancient biblical manuscripts amount to literary fossils of historical events—virtually exact copies of early history preserved accurately without embellishment. It might have been that historical research on biblical manuscripts would coalesce into concluding the Bible was an accurate and internally consistent account. But biblical historians instead concluded that every version of the Bible is a product of many human hands. Even so, isn't it possible those hands were guided by God? Historians generally answer that the early framers of the Bible were notably guided by a practical vision of what the church needed for survival and growth, by the pragmatics of religious practice. Additionally, over the centuries many official and unofficial revisions of the Bible have produced differing versions. There is no shortage of dispute about which of the multiple versions of the Bible in use is the most correct. Would a divine hand have guided several conflicting versions? The translation of various Bibles into over 2,000 languages adds further variation to scriptural meanings, some accidental, and others by the intentions of translators like Martin Luther (see Chapter 8). We must accept that the scope of human intervention went well beyond copy errors and modest changes during translation. The fiction of biblical inerrancy through divine guidance or inspiration is persistent enough to warrant a more detailed examination of the construction of the New Testament.

The main narratives collected into the New Testament were written at least half a century after the death of Jesus. Paul's letters may be the earliest New Testament writings, set down as early as 51 CE shortly before the Acts of the Apostles that were probably authored by Luke, a supporter of Paul and a rare Gentile among the disciples. The canonical Gospels represent stories and sayings attributed to Jesus by the unknown authors referred to as Mark, Matthew, Luke, and John who wrote about 70-90 CE (Tabor, 2006).

Throughout its formative stages in the 2nd century CE, Christianity was unable to settle on a fixed canon of beliefs. Because the early Christian churches were scattered about the Middle East, a medley of diverse views and teachings arose. Unification into a satisfactory New Testament took two centuries; it was not finished until the late 3rd century CE. The earliest surviving texts are themselves handwritten copies of copies, since the original scrolls disappeared long ago. The Chester Beatty papyri contain significant portions of the New Testament that date from about 200 CE. The Codex Vaticanus of the Vatican Library and the Codex Sinaiticus of the British Museum contain much more of the New Testament but were written after 300 CE. If Jesus himself left memoirs or other writings, they more than likely have crumbled and blended into the dust of the Holy Land. The bishops suppressed numerous unwanted writings, even denouncing them as heresies. It may surprise you to know that there were more than 30 early Gospels that didn't make the cut. There are complete copies of eight of these Gospels and another thirteen are partially preserved. But the entire texts have been lost for an additional thirteen—only their names remain. Even the intact Gospels came in several versions, like the four known versions of the Gospel of Thomas and the four known endings to the Gospel of Mark (Hedrick, 2002). During the 2nd century CE ecclesiastical officials selected the four canonical Gospels for inclusion in the New Testament. All of the remaining 30 or so were rejected. The Gospel of Mark was written around 70 CE and may have served as a source for both Luke and Matthew since 70% of Mark can be found within Luke and 80% within Matthew. John was written some 20 years after Mark (around 90 CE) and has little overlap with the other three. Before being judged acceptable, all four were almost surely edited to implement the vision of Christian church leaders. It was a proselytizing objective that required unquestioned acceptance of the bishops' canon of belief. The Gospel of Mark is generally believed to be the most historically accurate, even if it was distorted by "clarifications," editorial embellishments, and excisions. In Mark, Jesus is depicted as having an unexceptional childhood with no suggestion of the divine. Indeed, apart from adornment with impressive titles, Jesus made no explicit claims he was divine. It took many decades after his crucifixion for the notion of Jesus' divinity to gather enough momentum to gain widespread acceptance. Paul's depictions of Jesus as the "Son of God" should not be taken as a literal claim that he was divine, but rather that Jesus was empowered by the spirit of God.

In claiming direct succession from the apostles, the progression of bishops asserted their indisputable authority in matters of faith. Accounts of Paul laying on hands were used to legitimize the bishops' claim of their ultimate authority—that the bishops were the designated beneficiaries of apostolic succession. From succession it would arguably follow that communication with God, and ultimately personal Salvation, would require the intercession of accredited priests. Such was the bishops' thoughtful plan.

In assessing New Testament writings one can ask whether the editing was divinely inspired at every editorial choice point. How were the bishops informed about which Gospels were inauthentic and which were the authentic Word of God? Were all acts of text selection and editorial changes divinely inspired? How could any two theologians who disagreed over biblical inclusions both be divinely inspired? How was the decision to exclude the writings of other religious figures not an active suppression of additional words of God? Were the bishops divinely inspired in their efforts to destroy manuscripts they felt might undercut the authority of the church? Is the Apostles' Creed less divinely inspired than the most widely accepted creed in Christianity, the Nicene Creed? To address these issues it may help to take a closer look at the dismissal of Christianity's Gnostic wing (pronounced "nostic", from *gnosis*, Greek for knowledge).

Suppressed Texts: Dismissing the Gnostic Gospels

Subject to vigorous surveillance by church censors, most copies of the 30 rejected Gospels were tracked down and ultimately destroyed. For example, a copy of the suppressed Gospel of Thomas survived in Egypt's oldest Christian monastery until 370 CE when the Archbishop of Alexandria discovered it and promptly ordered it destroyed. What is known about the contents of suppressed documents largely comes from those that were hidden away in caves and other secret locations where they are occasionally discovered. In 1945 the Arab peasant Alī-al-Sammān discovered one particularly rich find of suppressed Gospels. He stumbled upon mysterious writings on papyrus stuffed into a three-foot tall clay jar in the dark recesses of one of the many caves near the village of Nag Hammadi in upper Egypt. Ever alert to biblical discoveries, historians soon realized that this simple earthen jar contained something for them far more precious than gold, the Gnostic Gospels—suppressed early Christian writings. From these leather-bound papyrus books, fifty-two texts ultimately survived a series of misadventures that began immediately after their discovery when a few books were inadvertently used as tinder in the Alī-al-Sammān family fireplace. What heresies did the Gnostic Gospels contain that warranted their suppression in the eyes of the bishops? The Gospel of Philip said "...Christ loved Mary more than all the disciples, and used to kiss her often on the mouth." The Gospel of Mary indicated that the disciples had agreed to accept the truth of Mary's teachings about Jesus. This clear elevation of Mary would have heightened the stature of women in general. Not only did powerful bishops suppress the Gospels of Philip and Mary, but they also suppressed the Gospel according to Thomas because, still more disagreeably, it placed Jesus closer to Mary than to Peter. The men who ran the church selected Peter as the designated successor to Jesus. Peter, not Mary, was to become the rock and founder of an all-male Catholic Church hierarchy.

By 180 CE Bishop Irenaeus of Lyons was zealously suppressing all Gospels as inauthentic and blasphemous, except for his favored four: Matthew, Mark, Luke, and John—today's canonical narratives of the New Testament. According to Irenaeus, it was the duty of all faithful Christians to accept without question the utterances of the bishops as the received word, including their claims of apostolic succession that marked the bishops as the sole recipients of the Gift of Truth. Opposed to this view were the Gnostics, a heterogeneous collection of Christian groups with diverse viewpoints and complex creation myths. In considering self-knowledge as divine, the Gnostic Gospels stressed that Jesus aimed to promote self-enlightenment and spiritual understanding, rather than offering Salvation through repentance of sin. It was neither good works nor faith in Jesus, but a nurturing of the internal divine spark that allowed the soul to survive bodily death. A similar view of self-reflection would re-emerge much later as the "Inner Light" of Quakers. The Gnostics believed that the god worshipped by the apostles was not the father of Jesus but was one of several inferior gods who were unable to eliminate evil, misery, and various calamities that befell suffering humans. Clearly, persons who found truth by inspecting their own souls were too self-reliant to be passively dependent upon the bishops and their closely calculated dogma. Irenaeus properly realized that any biblical Gospel that made self-examination pre-eminent would undercut the role and authority of the priesthood. Among the Gnostic writings, Irenaeus seemed particularly annoyed with the Gospel of Judas which he called "fictitious history" that painted the "traitor" Judas as a hero. It was not until 2005 that biblical scholars were able to translate a decaying copy of the Gospel of Judas and understand why Bishop Irenaeus found it so unsettling.

Box 6.1 The Gospel of Judas, 140-160 CE.

The recently published translation of the Gnostic Gospel of Judas deeply conflicts with traditional Christianity. Under unclear circumstances in the 1970s a scroll was discovered hidden near the Nile river delta in Egypt. Although an attempted 3 million dollar sale collapsed, the owner managed to get the expert clients to reveal that the scroll seemed both authentic and fascinating. Subsequently, the scroll sat for 17 years in a humid safe deposit box in New York State. Sold and resold, it continued to deteriorate into smaller and less legible fragments of papyrus, an early form of paper made from pressed plants. With the owners' indifference to the scroll's continued deterioration, its startling message almost disappeared into undecipherable bits. Eventually biblical scholars obtained the scroll's fragments and painstakingly pieced together the Gospel of Judas from 2001 to 2005. Radiocarbon assays suggest it was most likely translated between 220 CE and 340 CE from a 2nd century Greek original. (See the National Geographic magazine of May, 2006 for a pictorial account.)

To appreciate its heretical message one need only remember that orthodox Christians revile Judas Iscariot for his betrayal of Christ. In this sinister tale Judas took 30 pieces of silver to identify Christ with a kiss so the authorities knew he was the one to seize. The story of Judas increases our compassion for Jesus' suffering and unleashes our hostility toward those lacking loyalty and belief. Judas even became a convenient general symbol of treacherous Jews whom some Christians hold responsible for Jesus' death. The wholly different viewpoint of the Gospel of Judas is one of understanding and heroism. Judas was not a turncoat but Jesus' closest male confidant who courageously agreed to Jesus' request to identify him to the Romans. It asserts that Christ assigned Judas this role because he understood Jesus best. Apparently the other disciples had failed to understand the message of Jesus. Who but Judas, his most favored apostle, was better suited to "betray" him in order to accentuate his martyrdom? "You will exceed all of them. For you will sacrifice the man that clothes me.... [Although] you will be cursed."

Jesus seems to have provided private instruction for each of the disciples with the most apt pupils of his teaching being Judas and Mary Magdalene. (A recovered copy of the Gospel of Mary was first published in 1955). Since the Gospel of Judas was written decades after the death of Jesus, some scholars believe it intends to convey a political message of the failures of the apostolic leadership and the institutional church to live out the true message of Jesus.

Whatever its intentions, the Gospel of Judas will challenge contemplative Christians. It is jolting to accept Judas as an actual hero martyring his Master and later himself. Even more theologically numbing is the possibility that the basis for Salvation is self-understanding (gnosis) and discovery of one's divine spark rather than faith or good works. The threatening messages of the Gospel of Judas are not going to sit well with traditional Christianity. We may expect some alternative textual interpretations that demonize Judas once again, but first-rate biblical scholars have vouched for the more positive view of the Gospel of Judas. How many more revelations lie hidden in the caves of the Holy Land silently awaiting the chance to inform Christian belief?

The multiple Gospels in circulation reveal the wide diversity of early Christian viewpoints before the four, suitably revised, New Testament Gospels gained the upper hand. The Gnostic Gospels cast a skeptical light on major credos, including the virgin birth, the divinity of Christ, his resurrection, and the claim of a reserved and discrete relationship between Jesus with Mary. Details of the actual relationships between Jesus and the apostles are difficult to ascertain. Perhaps the canonical Gospels are more historically accurate since they were written several decades earlier than most of the Gnostic Gospels. Nonetheless, questions remain about which aspects of the various accounts are historically correct and which are expressions of human need and hope. There is no question that the clergy promoted their own role as essential intercessors between God and man. To ensure a monolithic canon that affirmed the preeminence of the bishops and to abolish any thought that Jesus had children whose divine male and female descendants might surface and checkmate the pope, the early church purged an assortment of threatening views, including those in the Gnostic Gospels. Surveillance to keep the Catholic canon free of contaminating heresies continues today.

Should The New Testament Be Strictly Embraced or Wisely Interpreted?

The New Testament has some coarse sections. Jesus said,

> If any man come to me, and hate (*miseo,* Greek) not his father, and mother, and wife, and children, and brethren, and sisters, yea, and his own life also, he cannot be my disciple. (*Luke* 14:26)

> And that servant, which knew his lord's will, and prepared not himself, neither did according to his will, shall be beaten with many stripes. (*Luke* 12:47)

> Except ye eat the flesh of the Son of man, and drink his blood, ye have no life in you. (*John* 6:53)

Today some might be inclined to wave off the last quotation as metaphorical, yet the Vatican maintains that the wafer and the wine at communion are literally transubstantiated into Christ's flesh and blood. In medieval times host desecration (scoffing at transubstantiation) was a serious offense; some individuals who failed to give due respect to the wafer as the body of Christ were tortured and killed.

A popular format for Sunday sermons is to quote a biblical passage to provide a scriptural seal of approval that sanctifies the minister's message. It is relatively easy to use a concordance, a detailed subject index, to find either a quote for peace or one for war because the Bible has many contradictory assertions. Even the same Gospel may provide conflicting quotes of Jesus. Thankfully some biblical contradictions can be explained as metaphorical constructions or a reflection of differing contexts. But many can't. In each of these three pairs which ought we to believe?

> The Lord is a man of war. (*Exodus* 15:3)
> The God of Peace. (*Romans* 15:33)

> Prepare war, wake up the mighty men, let all the men of war draw near; Let them come up: Beat your plowshares into swords, and your pruninghooks into spears: let the weak say, I am strong. (*Joel* 3:9-10)
> ...and they shall beat their swords into plowshares, and their spears into pruninghooks: nation shall not lift up sword against nation, neither shall they learn war any more. (*Isaiah* 2:4)

> And Jacob begat Joseph the husband of Mary.... (*Matthew* 1:16)
> And Jesus ...the son of Joseph, which was the son of Heli. (*Luke* 3:23)

So was the grandfather of Jesus Jacob or Heli?

Historical Accuracy and Guided Interpretation of the New Testament

On the trail of biblical meanings. It is a telling indication of human nature and need that the Holy Bible stands simultaneously as the world's most widely distributed, most widely accepted, and yet most widely disputed book. Its severest critics would say the Bible is a collection of narratives by many writers, cherry-picked and modified by its compilers to meet the needs of the bishops.

Even then, its frequently unclear wording encourages endless argumentation about proper interpretation.

As translators know, and as postmodernists have argued to a fault, the process of reading inevitably involves interpretation to grasp the intended meaning. In trying to understand what the writer had in mind, you hope that you share the connotations of each word and phrase. Biologists know that the word "theory" used in the phrase "the theory of evolution" means a grand synthesis that organizes innumerable data sets, regularities, and laws. The ordinary layman may use "theory" in its different meaning as a "notion" or a "guess" at an explanation. What a world-view of difference!

By admitting to "interpretation," the phrase "literal interpretation of scripture" lets the cat out of the bag. It is not unusual for two literalists to have different interpretations of the same scriptural passage. It is quite incorrect to believe the claim that devoutly read biblical verses have only one true, Simon-pure meaning. By saying, "Yea, the world is established; it shall never be moved," Psalm 93:1 was used to justify the heliocentric view of the universe, that Earth (the world) cannot move in an orbit around the sun. But the psalm doesn't exclude movement of the earth, only that Earth will not be moved by storms or earthquakes, or elephants or ... And look at the towering troubles that Revelation 20 has caused, spawning so many differing, yet powerfully life-changing views about the Second Coming of Jesus, rapture, dispensation, and 1,000 years of bliss or agony on Earth (see Chapter 16).

If biblical literalism is both historically and linguistically untenable, what does that say about the infallibility of the New Testament? More importantly, how might it change the impact of Jesus' words? Many who cling to literalism fear a slippery slope on which the reader begins to slide from a flawed Genesis into doubts about other recounted events, down into ever-deepening doubts about the accuracy of God speaking through scripture, ultimately bottoming out with the realization that as a human construction the Bible should be interpreted by humans according to their best judgment. As an evangelical Christian and professor of biology, Falk (2004) adopts a pragmatic strategy. He accepts everything in the Bible as correct, until proven wrong by science or perhaps by historical research on ancient manuscripts and the like. This approach is artificial in allowing no uncertainty and failing to come to terms with the Bible's conflicting and coarse moral regulations that bypass science.

Surely, we should take a stand in actively rejecting the horrific cruelty of the Old Testament, the Hebrew Bible, and the Qur'an. It is simply unacceptable to kill infidels (disbelievers) or apostates (those who have abandoned their earlier faith). And no praiseworthy society slaughters children. In telling it like it was, the Bible's harsh phrases reflect terribly brutal times. The New Testament passages advising when to punish slaves or how to press women into subservience are outmoded forms of brutality. Women should be free of abusive treatment and be able to enjoy all the rights available to others. (Sadly, as we will see in Chapter 8, this view is not shared by some Christian and Islamic leaders.)

Disagreements over the divine. All versions of the Bible reflect intensive winnowing and sifting to emphasize the more favorable phrases. As with the Old Testament there was extensive discussion about what to approve for the New Testament. For example, in 376 CE Bishop Athanasius of Alexandria compiled a list of works to be included in the New Testament. This list was then evaluated and altered by the Council of Hippo in 393 CE, and later, a modified New Testament text certified by the Council of Carthage in 397 CE.

Foremost among New Testament writings, the four canonical Gospels are frequently offered up as the literal Word of God. Yet, the work of many creative hands has shaped today's Gospels to fit the needs of the institutional church. After unknown authors set down their own written versions of oral traditions, major and minor changes in content were inserted before their adoption. Subsequently, down through the centuries numerous editors and translators added further changes.

To approximate the original narratives historians must work from fragmentary copies of copies of ancient manuscripts that themselves may have undergone extensive revisions from a lost original. As a consequence, markedly divergent accounts exist of the same events. In the absence of a true original manuscript as a gold standard, one should not be surprised when biblical scholars disagree vigorously over diverse textual sources. Intense arguments have arisen over the parentage and divinity of Jesus. Most wrenching has been the credibility of three claims: mother Mary's virgin birth, the divinity of Jesus and his biological relationship to God and spirits, and the resurrection or revival of Jesus from the dead, including his ascension into Heaven. The Mormons, Jehovah's Witnesses, Quakers, Unitarians, and some other denominations do not accept the concept of the Trinity (Father, Son, and Holy Ghost or Spirit). They have a point. The concept of Trinity somehow merge three Gods into one, for no Christian wants engage in polytheism with a separate God the Father, Jesus as God the Son, and God the Holy Spirit. The obvious resolution that God can appear in varied forms, such as Jesus, is an idea that gets skewered on the cross. If Jesus were but God taking the form of a man, then it must have been God who died on the cross. If so, how could a dead God have resurrected himself in the form of Jesus? If it was all a charade and Jesus a.k.a. God didn't really die because God is immortal, then this core tenet of Christianity is a hoax. It was no wonder the Catholic Church summarily declared it a heresy to state that the Trinity or the duo, Jesus and God, were crucified. How Jesus and God can be one yet meet separate fates remains a logical paradox. The Trinitarian idea of a split that is not a split is freighted with weighty paradoxes. These long-standing concerns are the subject of innumerable theological tracts whose dense opacity offers little clarity.

Others continue to question Jesus' divinity, or even whether Jesus actually lived. Was Jesus just a pastiche, a large montage, a medley of the lives of several itinerant preachers, embellished by the stock of stories and prophecies common in biblical times? Questions have also been raised about the accuracy of other biblical events. Perhaps as a consequence of skepticism, important doctrines have been modified over the years in order to strengthen participation and adherence. Key beliefs that have received more recent accentuation include the papal Declaration of the Immaculate Conception (Virgin Mary) in 1854, and The Assumption (ascension) of Mary's body and soul officially certified by the Catholic Church in 1950, nearly two thousand years after the claimed event.

The saga of the secret Gospel of Mark. It might be imagined that in compiling the New Testament, the bishops and the reviewing councils retained the historically verifiable. Yet, the great majority of the Gospels in circulation were excluded, even though aspects of some, like the Gospels of Mary, Philip, and Thomas, were quite informative. The Gospel of Thomas is a collection of 114 sayings attributed to Jesus. A majority of New Testament scholars believe that the Gospel of Thomas is a more accurate representation of the expressions of Jesus than the Gospel of John. The Gospel of John strongly reflects the programmatic concerns of the early church, especially the intention to trumpet the divinity of Jesus which the other Gospels fail to do (Borg, 1994).

Even the four chosen Gospels were selectively edited. Consider the extraordinary admission by Bishop Clement (ca. 150-215

CE) about his rationale for omitting a portion of the extended version of the Gospel of Mark that circulated in Alexandria. As background, in 1951 Professor Morton Smith discovered in the library of the ancient Greek Orthodox monastery of Mar Saba in Palestine a copy of a letter written by Bishop Clement of Alexandria. A bevy of modern biblical scholars agreed that Clement wrote this letter before 200 CE. In it he thanks an associate who played a part in removing offending portions of the Gospel of Mark. Some of Mark's writing seemed to advance the interests of a competing Christian sect, the Carpocratian Gnostics, reputed to be libertines who affirmed their cult by branding themselves behind their right ear. Clement's letter states,

> You did well in silencing the unspeakable teachings of the Carpocratians. ... For even if they should say something true, one who loves the truth should not, even so, agree with them. For not all true things are the truth, nor should that truth which seems true according to human opinions be preferred to the true truth, that according to the faith.

Writing with appalling self-incrimination, Bishop Clement explains the censorship of the Gospel of Mark before its inclusion in the Bible. Just as Mark selected those accounts of Jesus "he thought most useful for increasing the faith of those who were being instructed," so too Mark's own long account was altered by purging it of what Clement called the "Secret Gospel of Mark." Clement said that one should not admit, "that the secret Gospel is by Mark, but should even deny it on oath. For not all true things are to be said to all men." (Smith, 1974) Evidently, Bishop Clement believed that the Bible should dispense persuasion and be stripped, if necessary, of potentially damaging truth.

Controversy persists in this heated corner of biblical history (Hedrick, 2003; Brown, 2005; Jeffery, 2007). Perhaps Clement himself was duped by an altered text. Duped or not, it is probable that Bishop Clement was determined to shorten the long Gospel of Mark to fit the needs of the Church. The most general message to be absorbed from Secret Mark and other accounts about the Gnostic Gospels is that the compilation of the New Testament was the handiwork of robust human interventions.

Biblical Revisions: Surprisingly Diverse Motives

The text of the Bible has undergone numerous translations, as from Aramaic and Hebrew to Greek, Latin, French, Spanish, German, English, and ultimately over 2,000 languages. The biblical narratives that were collected and translated into English versions diverge in important ways from the older texts. What motivates efforts to "revise" the Bible? Is it to simplify the language, to weed out previous errors, and to update biblical scholarship with new information, or are there further objectives? Consider the frank commitment to the Received Word as stated by the dozens of editors directly responsible for the Revised Standard Version of the Holy Bible, copyrighted in 1946 by the National Council of Churches.

> The Bible is more than an historical document to be preserved. And it is more than a classic of English literature to be cherished and admired. It is a record of God's dealing with men, of God's revelation of himself and his will. ...The Bible carries its full message, not to those who regard it simply as a heritage of the past or [who] praise its literary style, but to those who read it that they may discern and understand God's word to men. ...and believe and obey his will.

The forceful closing statement suggests the editors had a mission beyond literature and historical scholarship. They aimed to render biblical writings into words that, in their most sincere belief, would represent the voiced authority of God. It is unclear how they knew when their efforts were successful. Was a majority vote sufficient to forward a revised passage as representing God's Word?

Has biblical diversity muddled God's Word? For much of the 20th century, traditional Bibles like the American Standard Version and the Revised Standard Version relied upon quaint, antique phrasing to animate ancient biblical scenes. Today, we are no longer limited to thumbing through one of these standard family Bibles to find a favorable passage—from a bookcase of Bibles we can select an entirely reworked one to get the messages we want to receive. The commercial success of the lavishly illustrated Bible called "Good News for Modern Man" (1966) helped to spawn dozens of popularized versions of the Bible. Four decades later Americans purchased more than 25 million Bibles in over 100 flavors. The English Standard Version (2001), Today's New International Version (2002), and the Holman Christian Standard Bible (2004) are examples with comprehensible phrasing and contemporary styling. Publishers increasingly emphasize Bibles with fundamentalist doctrinal messages in a vernacular that people want to read. For example, in the traditional New American Standard Bible *Exodus 34:14* says, "... the LORD, whose name is Jealous, is a jealous God." The knockoff Bible for fundamentalists, the "New Living Bible," has *Exodus 34:14* saying, "He is a God who is passionate about his relationships." This leaves us with a better feeling. Rick Warren's mega-bestseller *The Purpose-Driven Life* is a model of substituting Fundamentalist-Lite messages.

Everyone agrees that the Bible has wisdom and heartfelt and loving passages which have withstood the test of time. Nonetheless, there are instances of disagreement between biblical content and well-established modern factual understanding. These conflicts are best reconciled by treating disputed biblical passages either as metaphors or as misunderstandings prevalent in less informed times. Rather than pretend the Bible is perfectly accurate or fully consistent with modern knowledge, those who read the Bible with integrity will let go of passages in conflict with facts from physics, geology, chemistry, and biology. It is even more distressing to witness the painful attempts to bend scripture into prophesies of the future, sometimes by imagining the Bible was implanted with coded messages. The problem is not with the Bible, which offers some good advice, but with those determined to project their needs and ambitions into it. We can only wonder what the Bible might have looked like had its content not been so altered by the needs of the institutional church and the continuing desires of the clergy, the laity, and commercial publishers.

One might suppose that the richly textured expressions of modern conservative Christian doctrine are based entirely on biblical scripture. But the terms: original sin, Trinity, immaculate conception, Christmas, sermon, Catholic, pope, excommunication, infallibility, inerrancy, Palm Sunday, Last Supper, Eucharist, the Lord's Prayer, the Golden Rule, monogamy, and abortion appear nowhere in the Revised Standard Bible. Some of these familiar terms have ancient roots, but for others it took centuries after certification of the New Testament for them to be absorbed into the substance of Christian concerns. Evidently, human needs have added a layer of additional beliefs beyond scriptural messages.

Credibility, Slippery Slopes, and Metaphors

The Bible is widely studied because many passages tell us what we want to hear about spiritual life; it is revered because this scripture may be divinely inspired. If Genesis wrongly claimed that all species were simultaneously and recently created, how do you avoid a slippery slope of skepticism? First, as I have described, this slippery slope began not with modern skeptics of a finished product, but right at the historical outset from the way the Bible was selectively assembled as a subset of many narratives themselves subjected to editorial modifications of meaning. However, the existence of frequent passages suggesting probable human motives in the Bible's construction is no reason to become wholly incredulous about scriptural validity. Certainly, if the Bible is metaphorical, it means scriptural passages may fail to reveal God's messages with perfect clarity. You must choose how far to move toward skepticism. If a fairly considered biblical passage is consistent with the historical record, there is reason for belief. Biblical passages can be read as poetic reaffirmations that do not dictate spiritual beliefs.

In an 1893 encyclical Pope Leo XIII provided a formula for revising scriptural interpretations in light of scientific advances. He regarded scripture is unfailing inerrant whose discrepancies with science stemmed from early theological misinterpretations. An alternative fresh approach is to ask what form the New Testament would take if biblical events were occurring today, rather than 2,000 years ago. Assuredly many of the cultural particulars would change, like the travels of the apostles and their mode of expression. But if the verities are eternal, they should remain so today, even for the scientifically informed. Such an exercise is one way to distinguish the ephemeral from the everlasting in the Bible. From the present standing of women and from what is known of science and of biology in particular, it is unlikely such a 21st-century Bible would repeat the narrative that God cloned the first human female from the first male's fifth rib. Uncritical acceptance of the Bible as literal truth is self-deluding.

7

The Search for An Embracing God

Philosophical Questions About Origins and Purpose.

Deep questions about deep origins. Many fundamental questions have no ready answers. Why is there space and something occupying that space? Can space and its occupants arise where neither existed before? Can there be a measurable concept of "where" if there is no space, or "when" if there is no time? Why is there something rather than nothing at all? How could the universe start from nothing? If God created the universe, what created God?

And so we arrive at a consideration of the origins of God. The idea that the universe began as a creation of God hardly clears up the question of what was present beforehand. A God complex enough to create the universe is more powerful and more complex than anything we know. So it is a profound puzzle to grasp what mechanism could produce a complex God in the first place. There should be no ducking or dodging this issue that frames every effort to understand religious views of Creation. A series of evolutionary steps is a daunting option, but the best we can imagine for the creation of God himself. Theological responses to the origins of God like, "Goodness and divine creativity are insubstantive powers," seem unintelligibly evasive.

Searching for life's purpose. Why am I here on Earth, how

should I try to live my life, and what shall I have hope for? This raises still larger questions of meaning and purpose. What is the meaning or purpose of the universe—these unimaginable numbers of stars and planets sprinkled about in an incomprehensibly vast solitude? What is the meaning or purpose of life on Earth with its stunningly attractive, diverse, and at times treacherous plants and animals—most of them so small and well-hidden we will never even be aware of them? What is the meaning or purpose of my own life—my rather fleeting moments of existence on Earth? To assume God has complete understanding creates a conundrum about life's meaning. If God knows which specific events will occur in the future, then all events are already predestined. So why would there be any purpose in struggling to prove one can lead a virtuous life if the outcome is a forgone conclusion?

Does it also strike you as curious that questions about life's meaning or purpose are invariably framed in the singular? Do we know in advance that there is only one meaning to life on Earth, or just one purpose to our own lives? It seems that once again our needs have silently shaped the discussion. In this case we have a need to simplify. We want simple answers. Give us one overarching meaning for life, not a tapestry of complexly interwoven meanings, and certainly not a jumble of meanings. So we say, while impatiently shifting our feet before the Almighty, "Please give me the one true answer." We seem destined to construct the questions and also set limits on the answers according to our needs. Each human life tends to construct its own simplified meaning.

Initial Searches for God: Chance and Need

Stumbling onto a mini-god. Imagine you are out for a dawn stroll on a remote tropical beach. There on the high side of the beach is a large jade jar partially exposed by the surf. With difficulty you dig out the jar and pry off its gritty lid. Out struggles a grateful genie who says, "Whew, am I relieved. It's been a long time." "Then again," says the genie, "I can see from the hopeful look on your friend's face that I need to tell you about my limitations. I have never created life nor have I been able to alter the course of biological evolution. Still, I am so thrilled to escape confinement and to be able to enjoy this delightful world again that I will provide you and your friend with long and happy lives on Earth, to be followed by a carefree afterlife. Here, use this magical debit card—it will never expire." And so your friend sets off to pay his hotel bill. Later after a few years of trouble-free use, what has become of your friend's motivation to do good deeds to satisfy

God? After all, a steady stream of free earthly comforts seems like credible proof that he can also expect a happy afterlife as promised. Do we expect this delighted debit cardholder will continue to dutifully attend mass on Sundays where he had more than once prayed for a well-paying job? As noted earlier, priests are perfectly aware that the latté crowd is likely to drift away from religious practice if they are nestled into material comforts and confident in their future Salvation.

Some creative claims about God. According to theologians divine creativity includes whatever actions God has taken or will take in regard to life. Most theologians believe God had some role in the initiation of life or in its evolution. Process theology asserts that life and its diversity are the joint product of natural processes and divine influence. Divine influences in a stealth mode could glide under the scientific radar, especially if they were one-time events, since science depends heavily on repeatable demonstrations. No one can rule out the possibility that God has judiciously sprinkled benevolent mutations into the mix. Most mutations are harmful to the recipient. So if God did introduce novel genetic change, is he responsible only for the "good" outcomes?

Regularities in nature must depend upon physical and chemical principles that underlie the higher order consequences in biology and behavior. Residual mysteries, like unitary or episodic events, might reflect divine influence—but which ones would these be? Some are good and some are bad, depending upon your perspective. Eons ago God might have nudged an asteroid to hit the earth and eliminate the dominant dinosaurs blocking the way to the diversification of mammals. So the first problem is, how do we recognize which events reflect divine influence?. And the second problem is how was energy transferred from the supernatural realm to the natural realm? Even if it is only a transfer of information that affects a simple organism, the information transfer must alter some state of matter or energy. Even to control what an individual prophet says or writes, God would need to intervene in nature. There is no known mechanism allowing the supernatural to influence the natural and no credible evidence that it does—modern miracles are hard to verify.

A God without limitation would conflict with reality. For example, an all-loving God would be in conflict with an omnipotent God. This is so because if God were both all-loving and all-powerful, it is difficult to explain why God would permit suffering that harms the clearly innocent—unless we humans are being deluded.

Some attributes of God consistent with science. It is interesting to examine the long pauses in God's activities as understood by scientists who are Christian. Above all, God is astonishingly patient. Fourteen billion years ago God started the universe off with a Big Bang. Then for several billion years he passively let things settle down as dusty matter congealed into stars and planets while the universe expanded outward. About 3.5 billion years ago he might have created life on Earth, which biologists believe would have evolved without further divine direction. If you believe political leaders, God has intervened periodically in human affairs. Those who proclaimed themselves to be leaders anointed by God included tribal elders, chiefs, city and state leaders, popes and other clerics, kings, queens, emperors, and some politicians and dictators. God stirred himself to occasional action in war and peace, including some 2,000 years ago when he sent his only son Jesus down to Earth to be sacrificed for the cause of human Salvation. Only then did the work begin of answering prayers and admitting people to Heaven. The multi-billion year interval between the beginnings of single cells and the emergence of humans is perhaps the most curious gap in God's resume. God was certainly in no rush to trot out humans. We are distinct latecomers on Earth. Maybe God feared humans would annihilate many creatures and mess up the ecosystem. That concern seems warranted.

One of the great problem areas for conventional religion is meshing reality with our longing for a personal God. Many have wanted a God who created human beings as the most resplendent creatures in the universe. However, bolstered by a wide array of observations, the bright light of Darwin's theory of evolution has overwhelmed such vain rays of hope. Moreover, molecular genetics persuasively shows that while we may be the greatest of all apes, we are still apes by genetic composition and descent. If it weren't for this towering stack of scientific facts, humans would have a case that they were the specially created poster-children of God. Trying to kindle the God of one's dreams can be like a Boy Scout or Girl Scout persistently twirling a stick while patiently expecting flames to leap from the smoking cedar shavings. It takes strong religious faith to imagine fire where there is only smoke.

Certainly, honesty requires that we adjust our view of God whenever scientific discoveries require revision of historically claimed influences of the divine. The content of scripture while it was being written was unavoidably constrained by human understanding of the period. No one would expect to find any religion's ancient scripture couched in the language of modern biology with as yet unknown concepts like DNA and RNA, nucleotides, point mutations, amino acids, ribosomes, and numerous other terms needed to communicate a vastly enlarged understanding. Scripture is a spiritual guide that needs to be interpreted in light of current understanding.

Traditional Christians necessarily feel a sense of the mystical in the presence of their God. It was the Creation of man from dust and the breath of God along with the sacrifice of Jesus' life for our sins, followed by his resurrection and ascension that comprise some of the mystical events that may urge us to a personal relationship with God.

Some attributes expected of God. It is overwhelming to contemplate God as the First Cause. A God able to start the Big Bang 14 billion years ago is incomprehensibly powerful and ancient. If this God continues to act, it is a profound dilemma of our making to imagine how a God who is unfathomable in size and shape, power and age might take the form of Jesus and some 2,000 years later remain today as God and Jesus; the personal confidants and benefactors of billions. New cosmological understanding about the stupendous temporal and spatial scale of our universe makes it even more unlikely that God could be all of those things. Will humans ever make sense of such irreconcilable conceptions? The clergy think God will remain mysterious. Many don't even want to raise pointed questions.

Traditional Christian expectations of God. The Christian's preference is for an all-powerful God who ever since Creation has not only interceded in major crises, but personally continues to answer prayers from devout individuals. The Christian mainstream is eager for a God who can provide for our basic physical needs and creature comforts on Earth, mostly by keeping us from harm and pain and suffering, and offering a pleasant afterlife. In order that God be our Protector and Savior, we in turn should be humble, reverential, and virtuous. In addressing our spiritual concerns, God should also provide answers to our simple yet profound questions. Our core concerns are easy to describe. While we are alive we want to avoid adversity and after we die we want to go to Heaven or the equivalent. Briefly put, we say, "Please hear my prayerful request and help me now and forever." Where do we find the God we need? Such a God is all-knowing and all-powerful and is fully supportive. As we will see later on, this heartfelt desire may require more than physical principles and consistency can bear.

Those of us who want a God, desire more than a God who merely sets events in motion and tinkers so subtly that his actions are seldom evident. We want a loving, personal God who will intervene to protect us and our loved ones on Earth and provide for a Heavenly afterlife. Ideally we would have private communications with God. God may invisibly influence private thoughts and feelings not accessible to others. Sadly, there is no persuasive evidence that God communicates with individuals. Is there anyone among us who is gaining accurate predictions of the future which could be used to prove such personal revelations? Please step forward now. If God were to actively initiate communication with a person, it would require interactions between the supernatural and material realms.

Nonetheless, many people have a need to find God. And wherever such a need is backed up by irrepressibly strong feelings, God will emerge. If the feelings are so intense they overwhelm facts, then, for better or worse, the God of faith will have whatever properties the faithful wish to assign. There are astounding numbers of individuals who confidently claim to have found God and to understand his intentions, including his preference for their denomination. It is truly remarkable that so many people unreservedly claim to know so much about God—above all else they know that God provides them with personalized attention and concern and a plan. "I believe God gave Christians privileged standing over all other creatures and the majority of non-Christians. God created me. God looks like me. God speaks to me and answers my prayers. God protects me from harm. It is God who informs my meaning and purpose. I am God's instrument. My religion is God's true religion. God will guide me to Salvation and Heaven." Some of these may be true, but taken together this "I've got my dibs on God and a spot in Heaven" stance seems to be a collection of self-serving constructs in need of restraint and humility. A narcissistic devotion to an anxious ego can end up morphing God into a personal servant, instantly ready at our calling. In all of this there is a tendency to elbow others into terrible fates because "Heaven is reserved for people like me and my denomination." Which is worse, being a member of a private club that excludes Catholics and Jews, or belonging to a religious cult that excludes all non-fundamentalists from eternal life (Price, 2006)? Selfishly, some sects still claim that not one citizen of ancient Babylon, China, Egypt, Greece, Inca Peru, Mayan Central America, and many other venerable sites with accomplished civilizations ever had any opportunity for Salvation prior to Christ. Where is the fairness in allowing redemption only for Johnny-come-lately Christians, and even then only for the subset who are members in good standing of the favored denominations? This seems like a frontal violation of the Golden Rule. If you are understandably skeptical of such self-serving claims of privilege, what can be said about the true standing of humans with the divine? In exasperation some are ready to call off the entire God project. If you consider our current view of Thor, Zeus, Wotan, Poseidon, and other ancient top gods, it seems that "we are all atheists about most of the gods humanity has ever believed in. Some of us just go one god further (Dawkins, 2003, p. 150)." But search we shall.

The Reasoned Search for God

Reasoning is the first option for philosophers searching for God. One begins by trying to prove God exists. Stretching back thousands of years, many excellent minds have carried out reasoned searches for God—some determined individuals have spent their entire lifetime seeking God and God's attributes. Since modern theologians are generally skeptical that a personal God can be proven by rational analysis, we turn to much earlier times to review the efforts of two classical scholars who had the time and commitment to seek God through reasoned inquiry.

As the director of a major mosque in Baghdad, Abu Hamid al-Ghazzali (1058-1111) devoted himself to finding the Islamic God through reason. In words that resonate with the later emptiness of Mother Theresa, "I have poked into every dark recess, I have made an assault on every problem, I have plunged into every abyss. I have scrutinized the creed of every sect; I have tried to lay bare the innermost doctrines of every community. All this I have done that I might distinguish between true and false, between sound tradition and heretical innovation" (Watt, 1953, p. 20). Unpersuaded by the proofs of others and exhausted by his own failure to find convincing arguments, he lapsed into a deep depression that "shriveled my tongue until I was prevented from giving instruction" to students (Watt, p. 59). In order to recover, Al-Ghazzali spent a decade as a Sufist Muslim, a contemplative recluse. At the end of this period he concluded that a personal religious experience of God was the only way to move beyond the unsatisfactory reach of logic and reason.

Among Christian theologians, Thomas Aquinas (1225-1274) assembled five proofs of God's existence. His arguments were so well-regarded they succeeded in influencing Catholics and Protestants for more than four centuries, although today they are generally considered unpersuasive. In brief, Aquinas saw it as logically necessary to have a God as the Prime Mover who both set all things in motion and established the true beginning of the unbroken chain of causal events. (Pope John Paul II believed that these arguments rendered the ultimate origins of the universe unto God rather than to secular physics.) Aquinas drew strength from the intricacy, beauty, and functionality of many organisms and concluded that God must have been responsible for such intelligent designs. (The failed "Intelligent Design" movement is thoroughly reviewed in Chapter 15.) The existence of complex designs in biology remains a popular theological argument for the existence of God as the Creator. Who but God could have created organisms as complex and beautiful as, as...well, as us. If we accept that explanation, we must deal with the further question of God's origins. Because nature displays a hierarchy of sophistication and excellence, God must represent the pinnacle of perfection in the great chain of being according to Aquinas who did not mean to imply by this that God evolved from simpler forms. God came first. Nonetheless, how likely is it that right from the outset a swirl of energy and matter would happen to coalesce and congeal into a perfect and stupendously complex God? There is no doubt God would need to be stupendously complex in order to create life's sophisticated organisms and arrange them into an elaborate tapestry of millions of interacting microbes, plants, and animals. On Earth, biologists know that evolution was up to the task of ever so slowly creating the broad diversity of complex organisms and their interwoven life histories. Indeed, not only is evolution the sole scientific explanation we have for how life arose on Earth, an evolutionary process is also the best option for how God came into existence. Rather than forming spontaneously from scratch, it is more likely that God is the last most sophisticated product of a long lineage of complex beings. In spite of this, a lineage of God, Father God, Grandfather God, etc., goads theologians who much prefer to skip the question of God's origins. We await any credible non-evolutionary detailed explanation for the origin of God. What astounding process was responsible for this mega-miracle—this everything from nothing?

Today, many theologians, scientists, and philosophers find rational arguments for God's existence unpersuasive. Even Aquinas admitted that rational arguments for God paled in comparison to his intense feelings about God that were aroused through prayer and meditation. Even if reason and meditation lead one to believe God exists, it still remains to discover God's attributes. This is no easy task. To begin, what does God look like? Eastern Orthodox Christianity has accepted paintings and sculptures of God as a useful way to enhance contemplation. Such religious portraits, which may completely cover the vast interior walls of a Russian Orthodox Church, may help contemplative individuals transcend the mystery of God, but at the risk of worshipping and kneeling to the physical idol rather than to God. Are we simply being egotistical, worshipping an idol crafted to look like man?

Yet, what is truly known of God? What can we know of God? Rationalism has revealed little; mysticism has claimed to reveal more. Many find that even after tantalizingly mystical experiences, God remains remote. The rapture of divine intimacy comes most intensely to the suggestible (Armstrong, 1993). In any event, ever since ancient times most theologians have agreed that the greater part of a fully endowed God remains beyond human comprehension. The Greeks distinguished between God's inaccessible essence (ousia) and God's revealed activities or energies (energeia) like those we might experience in prayer. It is still the same today. At best what little we know of God is through presumed divine actions, including revelations about features of God that individuals may receive, rather than through rational analysis. The theologian Karl Barth (1886-1968) suggested that knowledge of God depends upon revelation provided by divine initiative, not by the efforts that humans make. This viewpoint conveniently accounts for the failed spiritual efforts of many humans to connect with God, although they may properly wonder why they have been ignored.

Homing in on God's Properties

How humans imagine God's properties. We want God to be sufficiently alike us that we can comfortably relate. Even as God's power and love inspire awe and reverence, at the same time we need to embrace God's thoughts and feelings and form. Is it surprising that for all branches of Christianity a man-like Jesus represents these ideal properties? The standard preference of Christians is also to imagine God the Father as paternal—an iconic image that is well-accepted in Eastern Orthodox Christianity. Imagine instead how you would feel praying to a host of fine wires swirled like spaghetti into a giant translucent box, or praying to an enormous, dazzling mobile of mirrors reflecting colored rays of light emitting diodes. Even at the highest levels of technical and artistic sophistication, a God remotely like these objects would be as unappealing as worshipping a car-sized iPhone®. We are scarcely inclined to imagine God as a giant electromechanical device or even an amorphous spirit but greatly inclined to consider God in our image. As the self-appointed highest life form on Earth, humans have repeatedly voted that God should resemble us. Does this mean that in an earlier evolutionary era before the rise of apes and humans (including Adam and Eve) that God would have taken on the form of the most advanced mammal of the time? Or still earlier that God would have resembled a unicellular organism during the 1-2 billion years single cells thrived before there were any multicellular plants or animals whatsoever? Perhaps if we traveled still further back into the mists of time God would dissipate into an inaccessible abstraction. As viewed by one wit, "If the triangles made a god, they would give him three sides" (Montesquieu, 1689-1755). In the absence of specific structural information, conceiving of God as an amorphous spirit might be the least presumptuous model.

Free will and the limits of God's power and understanding. In the reality of our own imperfect lives, we must concede many areas of ignorance and limitation. Imagine the hospital scene where the attending physician is saying, "Unhappily our medical staff has no remedy for your tragic disorder. We have done all that we can. I'm so sorry." In contrast to the limits of medicine, the Christian God of the New Testament, when taken literally, is all-knowing, all-powerful, and all-loving. Just feel the simplicity and comforting purity in these three immensely pleasing attributes—limitless knowledge, power, and love. Even so, this conception has also vexed Christians for centuries. Writing in *The Bondage of Will* Martin Luther declared God's absolute power both shatters free will and brings "bad news and the herald of terror" (Campbell, 2006). Here Luther was recognizing that if God is omnipotent, he is responsible for human evil; lacking free will, humans could not be held responsible for creating harm and suffering, if they have no free will to choose.

Without probing into God's intentions we can ask about the implications of free will in humans. If the guidance from Genesis and the bedrock assumption of legal systems are correct, humans are responsible for any actions taken with free will. This limits God correspondingly. If humans are free to make on the spot choices, God cannot have foreknowledge of what these free actions might be. Knowable, predictable actions are not free. By definition, if human choices are truly free, God is not controlling either the choices or their immediate impact. Evidently God is neither all-powerful nor all-knowing. Free will means that it is humans who can be, and perhaps should be, blamed for much evil. Many philosophers and scientists would dispute the existence of free will; thereby reviving God's potential to be virtually all-powerful and all-knowing although at the expense of being saddled with responsibility for the evil actions of the humans he created.

Suffering. In our need to have a powerful God we place virtually everything within his grasp. Disregarding human free will for the moment, what happens today on Earth seems to be exactly what God wants to happen. Nonetheless, for God to allow enormous suffering from diverse causes is in conflict with his power and limitless love and compassion—inconsistencies that every era of Christians has been compelled to reckon with. Suffering and unrelated human evil actions are elephantine theological dilemmas that put God's powers into a circus-like spotlight. We will devote all of Chapter 8 to these issues in examining whether human free will or principles of evolution release God from responsibility for human suffering.

Reaching God Through Prayer

Let's think about the capabilities you would like to see in God. You might long for a loving and all-powerful God who over time would eliminate suffering and evil. Is God's hand at the tiller of medical advances that now relieve so much suffering? Why can't God help out a little more to subdue additional illnesses like cancer and heart disease? Closer to home, would you want God to protect your friends and family from disasters? How often does God support our prayers? What is likely or even feasible in the matter of prayer?

Healing through intercessory prayer. Just the prospect of receiving vital medical treatment is likely to be encouraging. Nearly a third of the benefit from a typical medical intervention can be attributed to the patient's belief that the treatment will be beneficial. These so-called placebo effects are widely recognized in medical practice as contributing to better health. Placebo effects are revealed, for example, if patients report significant improvement, even though the medication was just a sugar pill (the placebo) that contained no drug treatment whatsoever. A positive belief in a treatment can contribute to a positive health outcome. Prayers and similar spiritual activities are welcomed at medical centers, if for no other reason than their placebo effect.

The vast majority of Americans believe that the healing power of prayer has helped a friend or relative recover from illness. To rule out placebo effects, most scientific studies of the health benefits of prayer examine intercessory prayer (prayers made on behalf of the patient by concerned individuals who don't tell the patient they are praying.) Properly carried out, a test of intercessory prayer would eliminate placebo effects. A proper study would also ensure that the patient was not given higher evaluations because the staff knew the patient had received prayers. Therefore, in a well-designed study neither the staff nor the patients know which patients receive outside prayers and which do not. If intercessory prayers seem effective, then something like divine intervention might be going on. If prayers don't seem beneficial, it could be that the tests were insensitive or the prayers were inappropriately carried out, or something else went wrong with the mechanics of the study. Some people get annoyed at the mere idea of testing the efficacy of prayer because they say God doesn't like to be evaluated, as if they personally know how God might react to research on improving patient recovery. The outcome of the most recent thorough study of the effectiveness of intercessory prayer turned out to be surprising. (Box 7.1).

Box 7.1 Intercessory Prayers for Heart Patients.

Each year more than 350,000 Americans receive coronary artery graft bypass surgery. Considering that 50% will have a major complication within a month after surgery, it is understandable that patients recovering from bypass operations receive numerous prayers. In a well-designed study, Benson et al. (2006) reported on the effectiveness of intercessory prayers given every day during the first two weeks of recovery by a daily average of 33 monks or nuns enlisted from the Silent Unity Protestants from Missouri and two Catholic orders (St Paul's Monastery in Minnesota and the Community of Teresian Carmelites in Massachusetts). A list of the first names and last initials of all patients was provided to the prayer givers. For researchers to be able to detect a 10% or greater change from the ordinarily expected 50% rate of major health complications, it was necessary to study 1800 bypass patients, one third of whom were randomly assigned to each of three groups. Two groups of patients knew only that they might or might not be prayed for by several concerned, devout Christians. In this main comparison the group of patients actually prayed for did not have a significantly lower incidence of complications than those who were not prayed for. The remaining group of 601 patients knew before surgery that they would be prayed for. Surprisingly they did significantly worse (59% had a major complication vs. 52% of the controls). Perhaps they were overconfident and less attentive to self-care. Or perhaps they became anxious at the thought their condition was viewed as serious enough to warrant prayer. Or some may have felt guilty and stressed-out for failing to show notable postoperative improvement, knowing that many prayers were being made on their behalf. Another recent study—the so-called MANTRA II investigation at nine medical centers—looked at the effectiveness of intercessory prayer in hundreds of cardiac patients (Krucoff et al., 2005). Once again distant prayers failed to improve any health outcomes. These two reports are the most sophisticated recent studies on the benefits of intercessory prayer for cardiac patients. Because there are many possible reasons for failure, negative results on their own do not settle scientific issues. Absence of proof is not proof of absence.

Health Benefits of Religion and Prayer

Religiousness and healthy behaviors. An increasing number of research studies aim to investigate the impact of religiousness on human health. Several published studies bolster the many anecdotal accounts linking certain religious practices to better health (Lee and Newberg, 2005). Churches that promote healthy lifestyles can claim some success. Some denominations successfully promote less frequent and less risky sexual behavior along with reductions in smoking and alcohol consumption, and improved nutritional habits. Yet, for many reasons, including the difficulty of measuring religiousness, it is difficult to provide persuasive evidence that health is boosted merely by increased personal religiousness. Are those who attend church more religious than those who rely on televised religious services? Do those who frequently pray and read the Bible have greater religiousness. In God's eyes is the merit or quality of religiousness the same across different religions and denominations? Unless one can measure religiousness, it is not possible to determine whether it improves health.

Is the Male God Aligned With the Concerns of Women?

Women and Islam. The rules set out in portions of the Bible and Qur'an clearly favor men. Advocacy of male dominance is an ancient theme in Judaism, Christianity, and Islam where it continues today. As Hashemi Rafsanjani, the ex-president of the Islamic Republic of Iran, put it,

> Equality does not take precedence over justice. Justice does not mean that all laws must be the same for men and women. One of the mistakes Westerners make is to forget this. The difference in stature, vitality, voice, development, muscular quality and physical strength of men and women show that men are more capable in all fields. Men's brains are bigger so men are more inclined to fight and women are more excitable. Men are inclined to reasoning and rationalism, while women have a fundamental tendency to be emotional. The tendency to protect is stronger in men, whereas most women like to be protected. Such differences affect the delegation of responsibilities, duties, and rights (quoted in Warraq, 2003, p. 203).

Rafsanjani's basic assertion is that unequal capabilities under God justify unequal treatment under the law.

The more general question is whether biological differences between males and females provide a rational basis for special privileges for either sex. Let's examine some of the differences in the standing of women, first in Islam and then in fundamentalist Christianity. (In the citations below, the numbering system for the Qur'an, but not the translations, generally follows that of Pickthall, 1930.)

Consider these Qur'anic declarations:

> A woman's solo testimony is not admissible in a court of law (II.282).
>
> In financial matters, a woman's testimony in court is worth half of a man's.
>
> "A male shall inherit twice as much as a female (IV.11)."
>
> "Men have authority over women because Allah has made the one superior to the other, and because [men]

> spend their wealth to maintain them. Good women are obedient. As for those from whom you fear disobedience, admonish them and send them to beds apart and beat them. Then if they obey you, take no further action against them. Allah is high, supreme (IV.34)."
>
> "A wife who disobeys her husband will go to Hell (LXVI.10)."

For a sense of the potentiality of a threatening Hell, consider this personal account of a Christian woman, recently married to a devout Muslim.

Box 7.2 Limits on Religious Questioning.

"My husband was the very example of good manners and values. I had never met a Christian (nor Jew nor Buddhist nor atheist) who even compared to him in this. He assured me that I would never be forced to wear hijab (head to toe covering) except in prayer... [However], it seems he had neglected to tell me one thing: that he had been allowed to marry me as a Christian, and as long as I remained Christian I could ask questions because I was an unbeliever anyway. But now that I had become Muslim, any doubt about anything the Prophet (Mohammed) did or said, any disrespect toward the Prophet, would make me an apostate. Apostasy, my husband said, would also automatically annul our marriage."

(Anonymous, in Warraq, 2003, p. 124-125)

According to policies in Islamic Iran:

Women are barred from becoming lawyers or judges. A woman may file for divorce only in extreme cases, such as the husband's insanity. For a man the declaration, "I divorce" is essentially sufficient. A divorced man receives custody of boys after the age of two and girls after the age of seven. If the woman remarries, she cannot keep even her young children. How does the Qur'an's tolerance of other religions compare with its stern treatment of women?

Armstrong (2000) views the Qur'an as fully accepting of other religions, stating it "recognizes the validity of all rightly guided religion, and praises all the great prophets of the past" (cited from Mohammed Asad, *The Message of the Qur'an*, 1980, Qur'an, 5:65, 22:40-43, 2:213-215). It also states that, "there shall be no coercion in matters of faith" (Qur'an, 2:256).

Nonetheless, the Qur'an also has some other sections that are not so accommodating. "The unbelievers are your inveterate enemies" (IV.101). "Unbelievers are those who say 'God is the Messiah, the son of Mary'...Unbelievers are those who say 'God is one of three'" (V.72-73). The Qur'an instructs Muslims to fight disbelievers (VIII.65, IX.123) and kill them (II.191, IX.5) or convert them until only Islamics remain (II.193). Being subjected to streams of boiling water that dissolve the flesh is one of several recommended gruesome punishments of disbelievers (XXII.19-22, LXXIV.26-27, XVIII.30).

Are we being misled by verses taken out of context? Well, according to Ali Sina, "The Qur'an is not organized by content. Verses are arranged in order of increasing length rather than according to the subject. Yet, the Qur'an has many verses that teach the killing of the unbelievers and tell how Allah will torture them after they die" (Warraq, 2003, p.148). Like the Bible, it appears that the Qur'an has contradictory declarations—both supporting and rejecting unbelievers.

The Qur'an permits unlimited sex with female slaves and with female captives of war (XXIII.6, XXXIII.50, 52; LXX.30). This is not merely license to enjoy the spoils of war. Rape is believed to defile the victim in the eyes of Allah and consign her to Hell. During the fighting in 1971 in what is now Bangladesh, Pakistani soldiers are reputed to have raped more than 100,000 Bengali women before killing them as disbelievers (Abul Kasem, in Warraq, 2003, p. 200). The righteous must be strong to carry out their carnal duties.

Women and Christianity. Western religion hardly has an egalitarian record in its treatment of women. The most notorious abuses of women by Christianity are evident in the horrors of the Inquisition and the Salem witch trials in America. It won't do to pass these off as the uninformed mistreatment of mentally ill individuals who were mistakenly diagnosed as being possessed by Satan. After all, didn't the similar incidence of schizophrenia among men potentially provide an equal number of male witches for burning?

Exodus 22:18 says, "Thou shalt not suffer a witch to live." The Roman Catholic Church spent nearly four centuries in avid pursuit of witches, from the 1300s to the 1700s. The bull *Summis desiderante* of 1486 of Pope Innocent VII sanctified the witch-hunting guide, *Malleus Maleficarum* written by the Dominican Inquisitors, Heinrich Kramer and Jacob Sprenger. This handbook described how to detect, torture, and kill witches. Among the many deficiencies of women that were pointed out, were the seven primary crimes of witches: causing impotence, or its opposite arousing passion in men, castrating men, changing men into beasts, using or supplying birth control, performing abortions, and committing infanticide. This list of primary crimes is notable for its focused condemnation of female sexual and reproductive matters, an attitude of sexual revulsion that resonates even today.

Under the rules of the Old Testament, women could be bought and sold for a price. A notorious case is David's cutting-edge display of male power when he purchased King Saul's daughter with the foreskins of 200 Philistine heathens. For the female sex, virginity before marriage was mandatory, after which one husband was the limit. For men, taking several wives was acceptable.

Several contemporary Protestant denominations continue to offer women less than full participation in church affairs. The Missouri Synod of the Lutheran Church still excludes women from voting on church business, even though women have voted in national political elections ever since the 19th amendment to the U.S. Constitution in 1920. The Bible is often an instrument of power. In 2000 the Southern Baptist Convention excluded women as pastors by citing *Timothy 2:9-14*: "I suffer not a woman to teach, nor to usurp authority over the man, but to be in silence." Pentecostal, Eastern Orthodox, and Seventh-Day Adventists also refuse to ordain women. There are no woman imams in Islam or woman rabbis in Orthodox Judaism. No woman can become a Catholic priest (or bishop, archbishop, cardinal, or pope).

Traditional Catholic doctrine is structured to make it doubtful that God might be a woman. It would take some remarkable genetic arrangements for a female God (Holy Spirit), lacking a Y chromosome, to have begotten the Virgin Mary with a male child possessing the vital Y chromosome. A female God would make the Trinity even more baffling since God the female and Jesus the male would somehow be One. We shall never pray to a hermaphroditic God.

Finding God: A Scientist's Testimony

In seeking a personal God, it may be helpful to trace out the steps of one molecular biologist's path that led him from atheism to evangelical Christianity (Collins, 2006). Here are the successive steps taken by Francis Collins toward belief in God through Jesus.

Step 1. Atheism seemed coldly indifferent—the cosmic void lacked personal warmth.

Step 2. A theistic (deistic) God may have impressively created an awe-inspiring universe some 14 billion years ago but his subsequent absence provides little comfort today.

Step 3. There is a present God who cares about our world because one can sense that humans have a universal morality—a natural moral law evident in persons reared in love. God has cared about life on Earth including the evolutionary emergence of humans.

Step 4. Achieving communication with God is the most difficult transition. This quest for divine reassurance is best satisfied by a personal relationship with God. Dialogs with God that yield direction about personal concerns require a belief in an emergent power if not a directly tangible one.

Step 5. The most secure path to talks with a personal God is through a historically confirmed Jesus when God became cloaked in a human form. Humble, loving, wise, and courageous, Jesus experienced human feelings and sufferings like yours and mine. "I have found the special revelation of God's nature in Jesus Christ to be an essential component of my faith" (Collins, 2006, p. 225). Embedded in Collins' faith set is the view, like John Paul II, that God has directed evolution, and that humans emerged as somehow more special than other creatures. He does have a point there. We are pretty clever in our domineering way.

Collins assumes that God has controlled the evolution of humans whom he granted free will. Yet, free will subtracts from God's powers to control further human evolution. It is through mate selection that free will would have its most direct and profound impact on human evolution. The human species will necessarily evolve in accord with the genetic mixtures generated by its matings. In many cultures it is up to each individual to choose their mate, or perhaps more accurately in western cultures "men propose while women dispose." The truism of "sexual selection" has profound implications. If the collective mating decisions of a society depend upon multiple free wills, then God cannot be controlling this core determinative feature of human evolution. To claim that God surreptitiously arranges each mating choice or fiddles with the resulting combinations of genes is to deny that humans are exercising free will. Theologians agree that when God gave humans free will, he ceded that fraction of his power to people. God might suddenly reclaim that power and convert humans to automatons, but under the present theological arrangements humans, not God, choose their mates. Wedding ceremonies may invoke God's blessing upon the newlyweds, but this is well after the couple has made their free choice. Not wishing to undercut God, a truly devout individual might consider having an arranged marriage. More experienced elders may stand a better chance of discerning God's preferences for the direction of human evolution.

On the social front Collins keenly desires harmony rather than continuing clashes between science and religion. He counsels us to disregard the flaws of belligerent religious figures and organized religion in order to focus on the virtues of proper faith. Since 'tolerance is a virtue,' "...you must conduct your own search for religious belief," he says (Collins, 2006 p. 225-227). I have some difficulty reconciling his declarations of tolerance with his concurrent commitment to the community of evangelical Christians so intent on converting us to the inner peace of being with Christ, however softly sold in his recent book (Collins, 2006). Whether it is an evangelical Christian, an evangelical Jew, or an evangelical Muslim who attempts to convert others to his faith, such zealotry is a declaration of intolerance of other faiths. A "tolerant Evangelist" is a contradiction.

What are we to make of all this? The central outcome of Collins' transformation is that he believes he has a personal relationship with God. He counsels others to transform themselves by absorbing the reassuring warmth of faith until they are aglow with contentment. At the same time many individuals have an imperative to get matters correct, even if that means living in a state less glowing but more knowing. They would rather stay true to their own sense of verifiable truth than saunter forth as a mellowed witness to the sanctity of faith.

Science relies upon critical analysis and logic. Arriving at conclusions by feeling is as much antithetical to science as requiring evidence-based proof is anathema to religious doctrine. These two different methodologies are incompatible. Scientists who are religious may deny leading compartmentalized lives, but they most certainly do not require scientific proof of their religious doctrines. In any event, if some religious scientists can live contentedly in the worlds of science and of religion, there is hope that religious citizens generally can learn how to accept science without demonizing its findings.

Abandoning God's True Religion

Western religions have traditionally threatened both men and women with severe punishment for apostasy, or the abandonment of one's faith. A rabbinical prayer of 100 CE read: "For apostates let there be no hope...let Christians and the sectarians perish in a moment. Let them be blotted out of the book of life" (Warraq. 2003, p. 15). To an outsider some acts of apostasy may seem rather minor. In Islam "any verbal denial of any principle of Muslim belief is considered apostasy. If one declares, for example, that the universe has always existed or that God has material substance, then one is an apostate. If one denies the unity of God or confesses to a belief in reincarnation, one is guilty of apostasy. Certain actions are also deemed acts of apostasy: for example, treating a copy of the Qur'an disrespectfully by burning it, or even soiling it. Some doctors of Islamic law claim that a Muslim becomes an apostate if he or she enters a church, worships an idol, or learns and practices magic. A Muslim becomes an apostate if he defames the Prophet's character, morals, or virtues, and denies Mohammed's prophethood" (Warraq, p. 16). The actual punishment for apostasy is a complex matter of Islamic law, depending upon the circumstances of the infraction. Immediate annulment of marriage and loss of property are not unusual. Muslim scholars agree that a convicted apostate must be put to death, but this punishment can be carried out in the afterlife. (From what we have been told, such a death may be preferable to an eternity in Hell.)

The treatment of apostates from Christianity varies with the denomination. The shunning of doubters is not unusual. One's rejection of official church doctrine may result in expulsion or excommunication with no hope of redemption. The traditional destination of a Christian apostate is Hell.

8

Evil and Suffering

Defining Good, Evil, Harm, and Suffering

It seems self-evident that the terms "good" and "evil" are neatly balanced opposites. However, most of us use the term "good" more broadly and loosely. In talking about good things, good events, good taste, good outcomes, and good feelings, "good" stands for "pleasant." I define "evil" more narrowly. Evil is restricted to "harm that is caused by someone's bad intentions," someone with mean-spirited or malicious motives. If we disregard the supernatural, only human beings are the cause of evil, generally by actions that are premeditated. Scary thunder and lightning, a hard fall on the ice, a cat bite, and even a bad dream, can be quite unpleasant experiences, but they are not evil because they lack human perpetrators. Nor is it evil if someone opens a door so quickly it accidentally bops you in the head. It may be careless or stupid and it may hurt, but lacking an intention to harm, it is not evil. Evil spirits are fanciful; they exist only in our nightmares.

Deeply regrettable, even horrible outcomes, are not necessarily evil. Few people would consider it evil for a policeman to kill a knife-wielding paranoid psychotic in self-defense. Of course, individuals bloated by feelings of power and righteousness may stretch self-defense to the point of attacking in anticipation. Moreover, according to some national leaders, it is not evil to declare a pre-emptive war that kills thousands of people, even when there is no credible evidence of a direct threat.

Although humans have much to account for, they are not responsible for most instances of harm. "Biological harm" embraces all forms of destruction or damage to organisms even to microbes and plants. To damage is to harm. Predators harm prey. Microbes and plants can be harmed, but lacking brains, they don't suffer. "Suffering" is pain, broadly defined, as it is experienced by humans and animals with suitably complex brains.

Humans have been responsible for frequent large-scale devastations. Harm and suffering were the outcomes, aside from whether these acts are considered to have been justified, or to be miscalculations, or accidents, or products of indifference. Fear has frequently led to tribal wars. The colonialization efforts of Spanish conquistadors and European settlers spread syphilis, measles, and smallpox that unintentionally killed thousands of Native Americans. In 1945 the United States intentionally obliterated Hiroshima and Nagasaki with atomic bombs. Industrial mishaps and pollution have harmed many lives, including over 2,000 individuals sickened or killed by methyl mercury-contaminated fish and shellfish at Minamata City on Kyushu Island, Japan. Humans typically try to justify the harm and suffering they cause.

Finding Justifications for Evil

Biblical times were packed with human evil. Certainly, the frequent sufferings of innocent minorities were excruciatingly apparent to the thousands of Christians who were publicly persecuted in the 2nd-4th centuries at the hands of the Romans—or sometimes at the claws and fangs of hungry carnivores in the Roman Coliseum.

For centuries ecclesiastical authorities justified killing heretics and "witches" in order to save their wretched souls from eternal damnation. Was it really necessary to torture children in particular? And if the torture was to get them to confess their preadolescent sins, why were they tortured even beyond their forced confession? Tortures like water-boarding and burning at the stake—the slow roasting of those who were most possessed of satanic heresies—were also useful in strengthening the Catholic Church by intimidating the populace into compliance with institutional goals. One of the important justifications for burning heretics is this remark of Jesus,

> If a man abide not in me, he is cast forth as a branch... and men cast them into the fire, and they are burned. (*John* 15:6)

At the peak of its enthusiasm in the 15th century the Inquisition was burning more than 50 women a day (Baigent and Leigh, 2000). It was not until the 1816 papal bull of Pope Pius VII that the Catholic Church officially renounced the use of torture. Regrettably, this welcome renunciation by the Catholic Church has failed to reach into the depths of modern governmental prisons. Various secular and religion-based governments, including the United States, shamelessly continue to kidnap, indefinitely imprison, and torture people.

What do the medieval Crusades, the torturing of medieval heretics, and the torturing of modern-day detainees have in common? Besides meeting the standard for evil (willfully inflicting premeditated excruciating pain to intimidate others and ensure bogus confessions from the generally innocent) the executive architects of torture claimed they were carrying out God's will. "Deus vult" or "God wills it" was the rallying cry in the first crusade in justification of cruel acts. President George W. Bush has repeatedly claimed his actions are informed by God's will. According to Amnesty International, and as admitted by President Bush in September of 2006, it remains the policy of his administration to subject United States prisoners to torture as normally defined. It is arrogant to justify torture by invoking the unverifiable and unassailable claim that God's will sanctifies one's actions. Leaders are at their most craven when they duck personal responsibility by shifting the blame to God. The implicit claim that torture is sanctioned from on high plays the ultimate trump card, escaping the restraints of human laws and the Bill of Rights of the United States Constitution. Whom would Jesus torture?

Personalized and De-personalized Evil

We are all too familiar with the abuses of cheats, crooks, thieves, and other scofflaws. Even generally law-abiding citizens manage to make mischief when they succumb to an excess of the seven deadly sins: greed, envy, gluttony, hatred, lust, pride, and sloth—each of them a sin of self-indulgence. However, in addition to scofflaws and the self-indulgent, there is a class of individuals doing harm who tend to get off lightly if they are even punished at all. In total denial of the impact of their actions, they think of themselves as plain, ordinary folks. After all, "mine is just like any other factory job—we just happen to make cigarettes that youths may smoke." Hannah Arendt unmasked this insidious harm with her famous phrase, "the banality of evil." Read on.

Nice people, perhaps you or I, can be responsible for hideous traumas, yet escape feelings of responsibility if our actions are quite remote in time and place from the ultimate evil outcomes. Let's trace out the story of how it is that ordinary Americans have helped to maim children like those in Cambodia. April 22, 2003 was a pleasant sunny day in northern California when picketers gathered in Sunnyvale to protest the manufacture of land mine components. As he entered the Lockheed Martin facilities for another day of work, one worker said defensively to the cluster of human rights supporters, "we are not doing the fighting; we are just doing our jobs." Meanwhile, continents away, the nursing staff at the Emergency Hospital in Battambang, Cambodia was completing another long day of surgeries on children maimed by land mines—hidden leftovers from earlier wars. Remnants of limb bones were trimmed of loose shards, and residual skin tugged over the raw stump and sutured. After healing during the months ahead, some of the older children would be fitted with orthopedic devices made at one of the 15 workshops for amputees located throughout Cambodia. With a prosthesis that approximates their stump size, children might then go to one of the 16 physical rehabilitation centers to be taught how to use their strapped-on device to hobble around or grasp at objects (Some had been playing with a "toy," as children will, when the land mine exploded in their hands). And in time they might visit one of the 130 physical therapists working in Cambodian hospitals. International relief agencies deserve high praise for their rehabilitative efforts in Cambodia. In 2001, the agency Handicap International Belgium distributed 335 orthopedic feet. Efforts to clear out the land mines are improving the situation. In 2001, 29,358 antipersonnel mines were found and destroyed in Cambodia. Undeterred by such carnage, Lockheed Martin, as the largest weapons manufacturer in America, continues to make land mine components. "Have a nice day at work, and you take care."

Concerning the Biblical Treatment of Evil and Suffering

The evil actions of humans. Yahweh, the God of the Old Testament, was tough, powerful, and brimming with vengeance. Human disobedience could bring a vile fate to violators of Yahweh's harshly enforced Mosaic laws. Subsequently, Jesus' remarkable love and compassion were pivotal in the progressive transformation of God into a gentle and benevolent figure. Equally important, as human suffering decreased, God's kindness increased because our conceptions of God inevitably track the human con life becomes even more benign, God will become corresp more loving and forgiving. Even then, in the dramas pl on our human stage we continue to seek relief from evil and suffering.

On the theological stage the vexing issue remains that evil and suffering exist at all. This vexation has created and animated theodicy, that huge body of ecclesiastical opinions that attempt to reconcile the desire for a protective, benevolent God with the evident presence of evil and suffering. Traditional Christianity treads down a torturous path to reconcile an all-loving God with human evil, since he created flawed humans at the outset. To review the momentous acquisition of original sin: God installed free will in Adam and Eve. Going against his command, they used this free will to enjoy an apple from the forbidden tree of knowledge of good and evil. This apple-eating episode caused God to banish Adam and Eve from the idyllic Garden of Eden and to brand all humanity with original sin and, no less consequentially, with mortal bodies. Original sin causes people to engage in self-indulgent excesses and sometimes do to evil by harming others. Further, with the possible exception of some saints, God placed a several-thousand-year hold on entering Heaven until Jesus became part of a complex plan to allow humans to get back into his grace if they asked him for forgiveness for their sins and pledged their faith in him. It is unclear why there was a span of some 75 generations before God sent Jesus to remove the "No Admittance" sign at the gates of Heaven. In any event, true believers in Jesus are to be granted a Heavenly everlasting life and might in the meantime expect to enjoy reduced suffering while on Earth. Even though this seems like a great offer on its own, God made the plan more appealing by sending his only Son Jesus to Earth to proclaim these rules and to heighten commitment by making the sinful feel guilty over his brief, tortured death. Guilt was generated because Jesus told his followers that as a direct consequence of human sinning, he, Jesus, would soon be crucified and die for our sins. Many children today are trained to feel personally guilty at the sight of a gaunt Jesus nailed to the cross some 2,000 years ago.

Free will and original sin are perhaps the most common theological explanations for human evil. If it is part of God's plan that we are all born with the gift of "free will" then moral failings are inevitable—and all the more of them, if we are tainted with original sin. By means of the toxic combination of free will and original sin the clergy have cunningly shifted the responsibility for evil unpleasantries from God to humans. To some minds it is also obvious that God needs to punish sinful disobedience in order to preserve his system of divine justice. Some also place blame on Lucifer or other evil forces God is unwilling or unable to control. Even so, is God really insulated from ultimate responsibility? After all, an all-knowing God surely also knew that a no-good serpent would be successful in tempting Eve to eat the apple. And it was God who gave people both free will and original sin, and withdrew immortality, or so we are told. Doesn't this also mean God made his son suffer on the cross for human sins and limitations which God himself permitted?

Natural calamities great and small. It is useful to separate the suffering caused by human actions from the suffering caused by natural calamities ("Acts of God" in legal jargon). Examples of relatively small-scale natural hardships include lightning striking a house, hail battering a cotton field, or an unavoidable minor accident or illness. There are also natural disasters on a stupendous scale like hurricanes, earthquakes, erupting volcanoes, huge lightning-triggered fires, and disease pandemics. Specific examples include the eruption of the Santorini volcano near Greece in about 1630 BCE that obliterated important Minoan settlements, the 79 CE eruption of Mount Vesuvius that destroyed Pompeii and Herculaneum, the deaths of many millions in the Black Plague in the 1300s and 1400s and in the flu pandemic of 1918, the drought that killed millions in Africa in 1993, the December, 2004 Indonesian tsunami that killed 230,000, and hurricane Katrina that flooded New Orleans in August 2005—the most destructive weather-related disaster in the history of the United States. How do we make sense of an all-loving God who allows so many natural disasters to inflict such widespread pain and suffering? Modern science views such destructive mega-events as part of the physical instability of the earth, its oceans, and its atmosphere. At least before the start of global warming brought about by human-produced carbon dioxide, no humans could be blamed for these large-scale calamities of nature that God failed to avert. It is hard to see how the physical calamities that bring harm to the lives of so many are in any way related to free will or original sin. As Rowan Williams, the Archbishop of Canterbury, put it in regard to the many thousands killed by the South Asian tsunami of Christmas 2004, "The question, 'how can you believe in God who permits suffering on this scale?' is therefore very much around at the moment, and it would be surprising if it weren't—indeed, it would be wrong if it weren't" (*Sunday Telegraph*, January 2, 2005, p. 22). But he then continues with the puzzling justification that since the world is characterized by the regularity of cause and effect, it would be odd for God to intervene in such natural calamitous events. That is a huge hole in the divine safety net. Perhaps, science and contemporary religion agree that we live on an active planet where large physical calamities do happen.

It can be a challenge to discern divine intent in life's many calamities or personally agonizing struggles. Although one can find misbehaviors that might remotely warrant retribution, like heroin addicts getting AIDS from contaminated needles, it is an even more mean-spirited claim that widely indiscriminate natural disasters are God's just punishment of evildoers. Yet, invariably religious conservatives deploy this lame excuse. Israel's Sephardic chief rabbi was among those religious leaders blaming the 2004 Christmas tsunami on God's vengeance against human sinning. He said, "This is an expression of God's great ire with the world."

Suffering of the innocent. It is theologically baffling why sinless individuals would be afflicted by disease, medical crises, or accidents. Some trot out original sin that taints us all. Moreover, suffering is not without benefits. Who can doubt that that suffering can build character and that the unpleasant features of life may motivate the otherwise self-satisfied to seek answers in God? But how can one justify the excruciating pains of the elderly whose character is already built up? It may also be correct that we come to appreciate good more fully through its contrast with evil. But does that explain why God allows colicky pain to torment newborn infants who have done nothing wrong? And what about individuals whose lives just seem to be dogged by a relentless succession of sufferings? They would be the first to tell you that fewer bouts of suffering would have been quite sufficient to stiffen their character. It is bizarrely contradictory that God would subject us to pain and suffering in order to re-kindle our belief in, of all things, his benevolence and goodwill. God as a merciless authority is the more likely lesson. And it seems a particularly thin justification that suffering is essential to sustain noble professions like charity, relief work, and emergency medicine.

A frequent justification for human suffering is that God is testing your faith. Does this ring true? Obviously this won't explain the suffering of anyone who promptly perishes in a house fire or other painful calamity. Suffering as a test of faith connects best with narratives like Abraham's willingness to sacrifice a son for a stern God who demands proof of loyalty—as if God didn't already know about Abraham.

The clergy are keen to help their flock cope with serious suffering which is unrelieved by prayer, lest the suffering prompt disbelief and apostasy. Those who can be persuaded that their trial is a test of loyalty to God are likely to remain loyal to the church. In thinking about the logic of a test of faith, you might wish to ask yourself how you would characterize a father who pushed his teenagers into an ice-fishing hole to test their loyalty to him. Isn't this like the theologian's claim that God-the-father arranges similar agonizing trials to test his children's' faith in him? (See Flynn, 2007, p. 657-660 for a skeptical view.)

For two millennia Christians have grappled with the paradox of an all-powerful and all-loving God who allows an abundance of earthly human suffering (Ehrman, 2008). As put in Martin Luther's shrill lamentation of 1545 "...as if it were really not enough for God that miserable sinners should be eternally lost through original sin, and oppressed with all kind of calamities through the law of the Ten Commandments, but God must add sorrow on sorrow, and even by the Gospel bring his wrath to bear." Luther's healing from this resentment of mankind's misery was aided when he intentionally added the word "alone" to his translation into German of the biblical scripture in *Romans* 3:28 that says, "... a man is justified by faith [alone]..." This simple addition to the Bible de-coupled Salvation from good works and succeeded in releasing Luther, and a stream of Protestants following after him, from guilt for sins perhaps too great for good works to quench. However successfully Luther's wording may have been in stilling the anguish of Protestants and smoothing their entry into Heaven, it certainly failed to abolish suffering and evil on Earth.

The theological basis for God's actions falls into two classes. Apparently, in numerous instances God has exact rules and judicial obligations that the spiritually privileged claim somehow to know. If we can find no sensible religious rule, as in unwarranted suffering of the innocent, then God must have reasons we can't fathom. Theological justifications for suffering will simply trail off with the dispiriting homily that "The Lord works in mysterious ways." This means that it is beyond the capacity of ordinary mortals to grasp the benevolence hidden in God's purpose. Or, more autocratically, "since God made us, God can break us in accordance with his unfathomable wisdom and plan." This sounds too much like an admission that in spite of its most sincere efforts, institutional Christianity provides no compelling understanding of why God's plan for life on Earth includes such widespread evil and suffering. Left this way, suffering is a serious shortcoming that lessens the likelihood of an all-powerful, all-good God.

Certain branches of Christianity propose a power standoff in which the good force of God and angels is opposed by the evil force of Satan and demonic helpers. Hence, human miseries reflect the ongoing battle between two bitter rivals, God and Satan. This relieves God of responsibility for suffering but regrettably sets limits on God's power to protect one from harm. God is no longer omnipotent but must struggle with a supernatural foe quite able to limit God's power to shield us from harm. It is a common belief that skulking about Earth are satanic spirits or demonic forces performing evil that God can't or won't eradicate. Worse still, some say that alien and evil forces can actually enter the human spirit, sometimes requiring removal by intense and risky rituals of exorcism. Judging by biblical estimations of the numbers of people sent to Heaven, compared with the multitude of non-Christians sent to Hell, Satan is more than three times as effective as God. Or are evangelical Christians wrong that non-Christians will be denied entry into Heaven?

Box 8.1 Satan a.k.a. the Devil, Mephistopheles, and Lucifer.

Historically, multiple "satans" arose first; only later did these forces coalesce into one supernatural being. At the outset disruptive satanic forces were sprinkled into traditional stories. "Satans" were diabolical or obstructionist processes brought into play by God (Pagels, 1995). As instructed by God, angels might disrupt undesirable human activities. (The Greek word *diabolos* means to throw something across one's path.) Satan has romped through Old Testament scripture as a trickster and tester of Christian resolve, as in "trying the patience of Job." After New Testament Christianity transformed God into a more loving figure, it necessarily transformed Satan into an independent evil being no longer acting on behalf of a provoked God but plotting against both God and mankind. When Lucifer, the angel of light, fell from God's grace, he was transformed into Satan, the angel of darkness. (See Milton's *Paradise Lost* for a poetic version.) As Jesus reportedly put it, "I saw Satan fall like lightning from Heaven" (*Luke* 10:18). Once personified, it became an irresistible human temptation to invoke Satan to scapegoat one's enemies as agents of the devil. Gnostics, Paulicians, Cathars, Jews, and Muslims were all portrayed as possessed by Satan and treated harshly by Christian powers. Such "spiritual warfare" continued even into the 20th century when Freemasons, immigrants, Catholics, anarchists, labor organizers, socialists and communists, and currently secular humanists, were depicted as agents of Satan. In this combat, dirty tricks against these satanically tainted people were permissible. The fear that children were in harm's way led to worried books with titles like "Teens and Devil Worship," and "The Devil's Web: Who is Stalking Your Children for Satan?" One antiabortion group claimed that Satan fed on aborted babies. While we reject the far out fantasy of a Satan who eats fetal flesh, those eager to stand up against the "forces of evil" must believe there is something to confront. It is wonderfully convenient to blame evil and suffering on Satan—after all doesn't he deserve it? Even today some live in fear of Satan as a sinister figure hatching plots against God and man. According to procedures that were revised in January of 1999, the Catholic Church continues to offer various levels of exorcism to rid minds of satanic infestations. By the 18th century, liberal Christian theology had concluded that Satan was mythical. In the modern enlightenment of the 21st century with our much deeper understanding we should demystify the origins of human suffering and mental illness. There are no grounds for prolonging the myth of Satan. It is time we parted ways. *Bye Bye Satan*, we don't need to trot out your myth as an explanation of evil or suffering or deranged thinking. Yet, since all of us have a personal stake in addressing evil and suffering, where do we place the blame if we don't blame Satan?

Ever eager to absolve God of responsibility, some Christian theologians view human suffering as an aspect of divine embrace. Don't you see, in God's plan humans can best appreciate the power and grace of God when God both endures suffering similar to ours in the form of Jesus on the Cross and then relieves our suffering through Salvation?

The tormenting presence of inexplicable human suffering on Earth has long invited inquiry. As the Greek philosopher Epicurus reasoned, when blameless unhappy events occur, it is because God either cannot or will not prevent them. For the devout individual, potent incentives to seek God include the occurrence of natural adversities, our suffering at the hands of others, and guilt over our personal sins. If instead, life on Earth were so idyllic it lacked these problems, even the prospect of an afterlife in Heaven might cause only the infirm and the elderly to take religion seriously. Biologists reason that some human actions traditionally classified as human sin reflect self-interest that has considerable survival value. After all, if there once were humans whose genes encouraged them to live selflessly without a thought or care for themselves and their kin, those humans are unlikely to have living descendants. Self-preservation requires some selfishness.

Animal survival and suffering. We should not neglect the plight of animals. If our understanding of biological history is correct, the forest primeval was never pristinely perfect. There never was a Paradise on Earth that might be spoiled by Adam and Eve. Studies of the natural history of animal lives have always revealed great suffering. Some were eaten alive, others starved. Routinely most offspring soon died in a massive apparent wastage—so many difficult lives full of pain and suffering under harsh and brutal conditions. The New Testament storyline was intended to be reassuring. Unfortunately, in the reality of life on Earth, matters are a bit more complicated. As the Creator of all life on Earth—and prescientific Christianity would have it no other way—God's creative actions also create complications. Clearly, for millions of years before humans emerged, there was great suffering among those of God's creatures whose nervous systems could experience hurt and pain. With their evolved requirement to eat protein, carnivorous predators, such as dinosaurs, have viewed smaller animals as food. Consuming one another is an essential aspect of the life histories of millions of organisms from microbes to mammals. The victims of these food fights often suffer before succumbing.

Those considering the widespread suffering of animals may be eager to find a way to absolve God of blame. Perhaps non-human organisms were also given the freedom to choose, and sadly, they chose to harm others. The suffering of their prey is the predator's responsibility, assuming tapeworms and their dim-witted ilk can be held accountable for their actions. In any event, many animals, including many parasites, are so specialized in what they can digest they have little choice in what to eat. Their anatomy and physiology limit them to a few sources of food. The lamprey's mouth only works when attached to flesh, like the side of a fish. An anteater can't subsist on twigs and leaves with its long snout and tongue, ineffective teeth, and the lack of a gastrointestinal tract suitable for processing plant material. Anteaters are specialists who ravage ant and termite nests. As a flightless bird without wings or a sturdy beak, the kiwi is unable to capture insects aloft or to utilize seeds. But with its long probing bill and a keen sense of smell, kiwis are highly specialized to stab earthworms. Perhaps the writhing of ants and earthworms involves no pain, only reflex-like behavior. Even though they seem in pain, it is not clear what they feel. But reptiles, birds, and mammals must feel pain when they feed on each other with their talons and teeth, plucking off pieces of babies snatched from their nests. In turn vertebrates are victims of the feeding habits of specialized parasites that suck blood from the skin or nutrients from the liver, or dine on succulent parts of the eye. It's a stretch to blame such cold-blooded parasites for their dreadful exercise of freedom of choice.

Despite prolonged theological efforts to blame mankind for all of human suffering, the burden has continued to rest at the feet of God. That is, until the mid-19th century when Darwin explained how complex and sophisticated organisms arose by the natural selection of heritable traits. What the theory of biological evolution has done is rescue God from responsibility for otherwise inexplicable suffering. Evolution explains what theodicy could not. Darwin showed why life histories have their unpleasant aspects. For example, cowbirds lay one large egg in the nests of other species. A newborn cowbird will heave out the other eggs or nestlings which soon perish on the ground. Why do they do this? Being rid of nest mates competing for food, cowbird ancestors that developed this merciless treatment survived better. As an evolutionary adaptation, harm is built into the life history of countless species. No individual microbe or plant suffers pain, but a sentient animal enduring "nature, red in tooth and claw" is another matter. Diverse complex organisms are the outcome of competitive struggles through deep time. Receptors that elicit a feeling of pain are needed for survival. Modern theological discussions of theodicy tend to wrestle with suffering's implications for God and mankind while largely neglecting the straightforward evolutionary basis of harm and suffering. The question for theologians has now shifted to why God uses or permits biological evolution.

Evolutionary analysis sees suffering not as desirable or undesirable but as inevitable. Pick any remote island and you will find various organisms that use each other as sources of food, because those that have done so have thrived. Animals need proteins, and only organisms produce them. A wholly loving God could have set an example like Buddha and spared animal lives by having all animals eat only plants. But this is not a plausible outcome of evolution. There is no paradise where all animals live solely on plant food and ignore the rich diversity of animal protein readily available around them.

The honest cleric would admit that God created co... that inevitably have led to animal suffering. So the question is: What do the clergy make of this? Unsatisfyingly, some claim that given God's goodness, he must have excellent reasons, even if we humans are too dim to understand his rationale. "God works in mysterious ways" is the inevitable phrase deployed to end the discussion. But it leaves us in the lurch about how we are supposed to behave. If the Lord intended that animals suffer, then should we be callously indifferent to the plight of endangered species like tigers and pandas, not bother to restock our winter bird feeders, fail to vaccinate our pets, or ignore a meowing or whimpering stray? Should animal shelters and humane societies fold their tents? No way. Humans are going to continue to extend compassion to animals.

The Biology of Torment: Accepting Life with Microbes and Pain

The actions of most of our biological tormenters are neither mysterious nor inexplicable. After all, the world is alive with a great variety of interdependent species that may interact as predators and prey. Happily, many interactions have benefited humans. The cellular mitochondria that supply our energy needs may well have been microbial parasites that long ago hitched a ride in our ancestral cells. We are also indebted to the chloroplasts that entered the ancestors of green plant cells permitting them to make oxygen as they use sunlight for power. Humans exploit yeasts to help make yogurt, wine, beer, and fluffy bread. Of course, yeasts have their own lifestyles; they may invade us and produce infections. The lifestyles of some rich and infamous microorganisms capitalize on people as warm treasure houses of delicious resources. These predatory lifestyles can drain us of vitality. But for the most part we have the built-in or learned weapons to combat microbial sicknesses. Public health sanitation and vaccines are now effective in preventing the ravages of cholera, typhoid fever, the Black Plague, and other illnesses that spread during the Dark Ages seemingly unchecked by chants and prayers. If pain is always unpleasant, should we look to the day when modern molecular medicine can permanently abolish it? 'Pain, pain, go away, and "never come another" day?'

Our skin is a mosaic of sensory nerve endings responsive to touch, warmth, cooling, and pain. There are many thousands of nerve fibers whose endings are specialized for carrying information about injury to the skin. These pain endings sport various receptors responsive to different kinds of intense stimulation that are damaging or nearly so. We have pain receptors keyed to detect burning heat, others for strong chemicals on the skin, or pinch, or for the chill of snow and ice. Cuts, abrasions, penetrating wounds, bruises, broken bones, even intense lights or sounds—all of these are detected by specialized pain receptors. Escaping from painful stimulation is an essential protective response if we are to minimize damage to our tissues.

Does the absence of sensory pain sound like bliss? Better reconsider. There are rare clinical cases of individuals born with little sensitivity to pain. Some lack pain as a result of a mutation that no longer allows sodium ions to enter and excite pain neurons (Cox et al., 2006). Imagine touching a hot stove and only realizing something was amiss when you smell your flesh burning—this is far too late. The skin of these rare pain-free children shows scars from numerous cuts and bruises they never felt. X-rays show their bones with multiple, poorly healed breaks they failed to notice at the moment of injury. A pain-free life will be a short, damaged life. The plight of these unusual individuals should make it absolutely clear that we need the sense of pain to warn us quickly of tissue-damaging stimulation. Pain is unpleasant, but a life without pain receptors is even more difficult and fleeting.

Microbial Diseases—Where Do We Place the Blame?

It is easy to empathize with the plight of early humans ravaged by childhood diseases and infections including plagues, and yet through all of those difficult times being wholly ignorant about microbiology and the causes of these calamities. With scant prevention and few remedies for sickness, it was easy for the ignorant and fearful to see causes in purely accidental correlations. It once seemed that sickness strode in with the cold night air, or was ushered in by a black cat, or an evil eye, or a suspicious event or visitor.

The combination of ignorance and illness gave medicine men the opportunity to test various natural products on sick members of the tribe. Even after spontaneous recoveries, it was the medicine man's powerful potions which received the credit. In actuality, experience did lead to many discoveries of specific plants as useful natural remedies. Even today, hospitals in China have two pharmacies, often side by side. One stocks modern drugs—the usual pills and solutions in familiar bottles and vials. The other pharmacy is for traditional Chinese medicine. As remedies for particular disorders, it offers plant leaves and roots and some dried parts of exotic animals, perhaps ground rhinoceros horn for impotence.

Disease and repeated bouts of sickness motivated our primitive ancestors to blame such torments upon mysterious evil forces that needed to be placated or even attacked outright. One popular plan was to kill an animal or a vulnerable person to appease a mean-spirited God who hopefully would draw satisfaction from such placating sacrifices and end the torment. Ideally the problem would disappear along with the sacrificial victim. Typical human scapegoats (sic) were either desirable (virgins) or expendable (criminals, captured members of weaker tribes, the mentally ill, and the elderly). Even if conditions failed to improve, the ritual sacrifice managed to intimidate the general populace into obeying those in power. One key to credibility was to hold sacrifices often enough that by luck a sacrifice would coincide with natural relief from a plague of locusts, drought, or some other quickly terminated calamity. Sacrifice today, problem gone tomorrow. So there you have it—palpable evidence that the sacrifice worked and cured the problem. If a sacrifice failed to produce relief, it could be argued that the situation called for still more sacrifices. It even made sense to use a pre-emptive sacrifice to ward off trouble in advance. In recent decades the task of overcoming plague and pestilence has shifted from the hands of spiritual professionals into the hands of health professionals. Let's see what kind of a job they have done.

The Success of Public Health Measures in Stanching Suffering.

"Thank your lucky stars." "Count your blessings." "We have been most fortunate." "We are truly blessed." These are the kinds of classical folk sayings which have attributed recovery from an illness to luck or divine intervention. In ancient Rome what passed for elite health care was downright primitive compared with Roman accomplishments in civil engineering. The best available pre-scientific medicine in Rome failed to save the lives of 12 of the 14 children born to the wife of Emperor Augustus Caesar. Nowadays, rather than trust our health to conjurers or supernatural powers, we seek out skilled care from those properly trained in medicine, nursing, and public health. Informed by basic science, today's health professions deserve credit for fantastic improvements. They have replaced Roman supplications to the gods and medieval medical quackery with treatments derived from scientific findings.

For many people, pre-scientific brutish times offered little in the way of creature comforts like warm clothing and shelter, safety and health, or security and freedom. The most important threat to one's health was inadequate sanitation in towns and villages. We now have high expectations for family health. So how is it that in modern developed countries most children are models of health and vigor, living their first decade with not much more than occasional sniffles and coughs, and bumps and scrapes? Improved sanitation heads the list of benefits—especially clean drinking water that helps to eliminate devastating diseases like cholera, dysentery, and typhoid fever. Second, improved nutrition prevents vitamin deficiency diseases like scurvy, beriberi, and pellagra. Uppermost have been the technical achievements in immunization that avert a slew of the infectious diseases that plagued previous generations. Turn the clock back to child health some 75 years ago when there was a significant risk of deadly childhood diseases. German measles, whooping cough, small pox, polio, diphtheria, undulant fever, typhoid, tetanus, rheumatic fever, tuberculosis, and influenza—many of these illnesses were severe enough to kill American children by the tens of thousands. Today's parents scarcely give a second thought that their children might contract any of these viral and bacterial diseases. Where have most of these terrible infectious diseases gone? Are the bugs still here or did they disappear with our obsessive habits of cleaning and washing? Or perhaps they were banished by the fervent prayers of Christian Scientists or Jehovah's Witnesses who would rather entrust their children's health to the prospects of divine intervention than to medical science. More rationally, we thank Alexander Fleming for penicillin, Jonas Salk and Albert Sabin for polio vaccine, among many contributors to medicine. Here are two more.

Alfred Elliot—a voice of courage. When Dr. Alfred Elliot was an Assistant Professor of Zoology at Bemidji State University in Minnesota in the 1940s he campaigned for the pasteurization of milk to prevent undulant fever caused by brucellosis bacteria. The flu-like symptoms of undulant fever began slowly, lasted for weeks, and then subsided only to repeat the cycle. The medical facts were clear, heating milk to 143°F for only 30 minutes would kill brucellosis, tuberculosis, and other dangerous microorganisms. The challenge was to overcome the farmers' fear that consumers might view pasteurization as tampering with a natural organic food. Does this sound familiar? Gaining consumer acceptance for the brief heating of milk was aided by using the term "pasteurization" which capitalized on worldwide admiration of Louis Pasteur who used heat to protect products like wine from microbial contaminants. During the tense public health campaign needed to pass laws for the pasteurization of milk in Minnesota, Dr. Elliot shielded his children from angry callers by instructing them not to answer the telephone. After a victorious campaign, pasteurized milk became the new standard, with healthier children the result.

Maurice Hilleman: A Hero Among Heroes. From the patient's perspective, prevention is the unsung ideal in medicine. Where possible, spare us invasive surgical heroics and wretched recoveries. Quiet prevention would be just fine. Having been vaccinated, you are unlikely to notice the vital moments when your newly marshaled antibodies rally to thwart microbial killers as they slither in through the portals of your body. The cryptic victories of vaccination against the microbial scourges of mankind have required many skilled hands, none more important than Maurice Hilleman (1920-2005). His dedication and initiative produced the vaccines that conquered measles, mumps, chickenpox, bacterial meningitis, pandemic flu, hepatitis B, German measles, Japanese encephalitis, diphtheria, whooping cough, tetanus and more! Today these vaccines quietly and unobtrusively save the lives of over a million American children a year. Take the history of measles. As recently as the 1950s in the United States four million children developed measles and some 3,000 died—every year.

About 50 years ago Hilleman's attention was caught by a report that 250,000 people in Hong Kong had come down with the flu (New York Times, April 17, 1957). As he read that the children were described as "glassy-eyed" he vividly recalled the 1918 pandemic flu that killed 20 million people. Within an hour he had ordered his team at The Walter Reed Army Medical Hospital in Washington, DC, to "get some samples of the flu virus from Hong Kong right away!" Stalled by the army's usual bureaucratic lethargy, the throat swabs arrived after a month's delay. Then, even working 14 hours a day it took an additional nine days to culture the Hong Kong flu virus. The results of a crucial test brought disturbing news; the blood samples of ordinary citizens had no antibodies against this strain of flu virus. This meant the Hong Kong flu was a full-blown threat. Unchecked by natural immunity, it could sicken millions. When the virus made its way to the shores of America by ship, as it surely would, it could feast upon defenseless Americans, just as the introduction of the measles virus and smallpox by the Spanish conquistadors had earlier decimated Indian tribes throughout much of the Americas. Judging from the 1918 flu epidemic, the United States was facing the potential of more than one million deaths. It was essential to make a live vaccine strong enough to cause a person to produce antibodies but not so strong as to produce full-blown flu. In weakening the cultured virus by transferring it from egg to egg, Merck Pharmaceutical used 150,000 chicken eggs a day. By the fall of 1957 enough vaccine had been made to immunize 40 million Americans. Even after a successful campaign of vaccinations the Hong Kong Flu of 1957 killed 69,000 Americans—but at least not the one million feared. Some time later, out of gratitude, Hilleman developed a vaccine against another virus responsible for Marek's disease that was killing chickens by the thousands. "I figured I owed it to the chickens," he said with a wry smile.

Hilleman's sense of practicality and broad vision seemed to mirror a childhood growing up on the range in Montana. A strict Lutheran upbringing helped cement his bedrock belief that "science ought to produce something useful" in return for public support. It also seems to have solidified the determination in his character, for he remembered with amusement how he defended himself when the minister caught him reading Darwin's *The Origin of Species*—in church. The twelve-year-old Maurice responded, "This book belongs to the public library, and I will report you if you take it away from me." After high school he was barely able to afford college tuition—all of $45 per semester! Eventually Hilleman won scholarship support that allowed him to get a Ph.D. in microbiology from the University of Chicago.

9

Touring Heaven

Heaven Looks Swell Compared to Hell

Out-of-body experiences. Naturally, everyone prefers the optimistic view that there is a pleasant life-after-death as offered by favorable scriptural passages and the encouragement of religious leaders. There are anecdotal reports of ascension-like experiences from those who have been near death. Surgical patients may report an out-of-body experience—feeling like they floated up to the ceiling and looked down on their body and others around them. Has their soul temporarily left their anesthetized body? While some of these cases might represent the spirit temporarily departing from the corporal body, more natural explanations of an altered body image are reasonable. Instances of patients recalling the surgical team's chatter in the operating room suggest these reports may reflect intervals when a lightly anesthetized patient was in a twilight zone of perception conducive to an out-of-body experience. We await cases where the subject or patient divulges some breaking news that could only have been gained during the travels of their disembodied soul.

In the clinic, stimulation of certain brain areas in an awake patient can elicit an out-of-body experience, or a feeling of another body close by, or a sense their body stands before them (Blanke and Arzy, 2005). In laboratory settings the combined influence of touch and vision can make a normal subject sense his body is a few feet in front of where he actually stands. This effect occurs if the subject's chest is stroked while simultaneously his video-camera eyeglasses reveal his displaced body image being stroked in the same way (Ehrsson, 2007; Lenggenhager et al., 2007).

The basic after-death alternatives. It requires no preview tour to be enticed by offers of a perpetual and carefree Heavenly existence. For the somewhat skeptical there ought to be no penalty to imagine there might be such a Heaven—assuming that caution doesn't offend God. To boost the incentive for firmer belief, some clerics point out that willful disobedience or disbelief risks a catastrophic fate. It can be an effective motivator to tell folks, "Remember the options, you will either go to a sublime Heaven or to Hell with its fire and brimstone." In some mythologies sinners are not actually be banished to a place like Hell, but instead may expect to be miserably reincarnated into some type of animal that lives a dog's life or worse.

Since they are some common features in primitive mythologies, frightful underworlds such as Hell must have deep roots. The Christian Hell was invented well before Dante's famous tortured visions appeared in his *Inferno* (ca. 1308 CE). Today, except as motivational carryovers of ancient religious traditions, it is only in our nightmares that we would imagine a divinity cruel enough to banish us to a place like Hell for an eternity.

Disbelievers have concluded that both Heaven and Hell are fictional—that death is the end of it. Your spiritual life ends with your physical death. Naturally, this alternative of non-existence is distinctly unappealing to those eager to continue to lead pleasant lives. The bleak alternative of oblivion prompts us to scout around for a better option, even if it is only to extend our lives by some years.

Admission into Heaven

Biblical stories about Jesus tell of a Second Coming, an imminent return to Earth. Before the resurrected Jesus ascended into Heaven he stated that he would shortly return to Earth in order to lead the faithful into Heaven. However, keen observers, including Paul and the other disciples, failed to observe this joyous event in the decades following Jesus' life. Two thousand years later we are still awaiting his oft-predicted return. There is always hope it will be in our lifetime. It would certainly be an amazing affirmation of faith were he to return.

The Bible indicates that the original sin of almost all humans, who lived in the six millennia before Jesus, disqualified them from entering Heaven. (Their fates remain unknown.) Practically speaking, we can be thankful that Adam and Eve were prevented from eating from the tree of everlasting life. One bite might have meant that some 20 billion of their immortal descendants would be populating today's world, which is already overextended in attempting to provide resources equitably to more than 6 billion mortal inhabitants. Graciously, after Adam and Eve were exiled, they merely died of old age, rather than being banished to Hell for eternity.

Although thoroughly outnumbered, the early Christians felt privileged. Around 55 CE Paul noted that there were just two categories of people; those few who, by being "born again" in baptism, had a guaranteed "citizenship in Heaven" and all the others—infidels—to be tormented forever by Satan and his demons. Christian baptism would secure a place in Heaven by protecting one from the clutches of demonic forces and by inoculating one against the perverse influence of secular human authorities.

In remarks that resonated with his followers, Jesus famously said that only through him could one enter the kingdom of Heaven. But he also said that as the King of the Jews he had come to save the Jews, not the Gentiles. Yet, mainstream Judaism never accepted Jesus Christ as the Messiah or Savior whose appearance had long been prophesied. Moses remained their mainstay. It was largely through Paul's travels that increasing numbers of scattered groups of Gentiles adopted Christianity as their religious faith, and Jesus Christ, a Jew, as their leader. Paul's peerless contribution to Christianity was to bring the Gentiles into the church. It stands as one of the greatest ironies of religious leadership that over time the teachings of Jesus were rejected by nearly all Jews whom he came to save, yet were accepted by the Gentiles whom he was prepared to reject.

Today, Jesus and his disciples, in collaboration with angels and saints, are said to be judging the men, women and children of all ages who queue up before Heaven's gates. It is a tough job. To develop a better understanding of our prospects for Heaven and its lifestyle, it is worth thinking about the probable admission procedures.

Who Will Enter Heaven?

Most people believe it is now possible to enter Heaven even before Jesus returns to Earth. Conceivably each of us is going to Heaven, regardless of how we live out our lives on Earth. Even so, in competing for adherents each sect presents itself as the keeper of privileged entry standards. Beyond loyalty to your mother church, your Salvation may require faith and grace, or good works, or a life with few mortal sins. The lowest admission standard is that faith alone is sufficient to guarantee admission. Essentially this requires that you have been born again through baptism and adopt Jesus Christ as your Savior. However, the antagonism that was evident among the apostles of the New Testament over the sufficiency of faith vs. the necessity of good works continues to be played out to this day in Christian services. Billy Graham's revivals have had a gracious and easy cadence where one could rise and step forward to be saved by grace. Then again, many other Christians, particularly Catholics, believe a life of virtue and good works is quite helpful, if not required. Much of mainstream Protestant theology emphasizes creating a heaven on Earth by devoted efforts and good works. Reminiscent of the Gnostics, a minority of Christians believes that personal understanding and spiritual insights are most important.

The Location and Properties of Heaven

What can be said about the location, ecology, and sociology of a sustainable Heaven? The clergy are characteristically vague about Heaven's physical properties. There are formidable difficulties in identifying, reaching, and organizing an idyllic place called Heaven. We take for granted many of the features that contribute to a comfortable and satisfying life on Earth. These include a warm place to sleep, several tasty meals a day, a pain-free body, a brain that works well, conversation and companionship, games and hobbies, enjoyable diversions like books, music, dancing, the Internet, and television, and perhaps the chance to tend gardens splashed with flowers. We have grown accustomed to modern creature comforts like fast food, snappy clothes, an easy chair, central heating, cars, cell phones, CD and DVD players, and so forth. It's interesting to consider whether upon arrival in Heaven we will be free to indulge in such pleasures and more. How will Heaven be organized and governed? Will every resident enjoy the liberty and the freedom to pursue personal interests? Where is Heaven located that it can offer us such freedoms and pleasantries?

Heaven has traditionally been placed somewhere up in the clouds. The exact location is inconsequential if we are only considering souls. But if we ascend body and soul, as traditional Christianity asserts (Haught, 2007), then the exact location of Heaven matters a great deal. It is inescapable that resurrected human bodies would have numerous physiological needs. In order to survive and enjoy life's many physical pleasures, our tissues and organs would need oxygen, food, water, and moderate temperatures.

It is one thing to fancy vast journeys of an imagined soul. But it is quite another to couple this to consciousness. The notions that the conscious mind can travel through the void of space, confer with other minds and last indefinitely is not supported by neuroscience. Being a feature of brain function, consciousness always travels with the brain and disappears when the brain perishes.

As experience with the defunct Biosphere projects near Tucson, Arizona has shown, there are daunting human and technical challenges to developing a closed system which is forever self-sustaining, managing to recycle what it uses and grow everything it needs. The immortal might find some tight requirements that make Heaven more restrictive than Earth. In the personal realm it is unclear whether you could import your CDs to Heaven or update your songs. Can you get the latest game players? Perhaps, instead of these smaller details we should turn our focus to the larger picture.

In looking beyond the clouds above for an accommodating site for Heaven, astronomers are on the verge of identifying earth-like planets that orbit other stars (Gaidos et al., 2007), although we must remember that every one will be light-years away from us. Accordingly, it might be more reasonable to inspect our own solar system for moons or planets possibly hospitable to our carbon and water-based life. Mars has some signs of water (Malin et al., 2006), but its surface temperature is 63°C below freezing. Regrettably, there is little chance of finding mammals, creatures like us, which would need more oxygen than is available in the thin atmosphere of Mars (only 1% of our atmospheric pressure). Europa, the largest of Jupiter's moons has an icy surface covered by clouds and an ice layer hundreds of feet thick. Titan, as one of the moons of Saturn, had been considered perhaps the prime nearby site that might be suitable for life—at least for some hardy bacteria. According to recent satellite assessments reported during the summer of 2006, there is a steady freezing drizzle from an atmosphere containing methane, ethane and nitrogen. Enceladus, a tiny frozen moon of Saturn, has some geysers that may or may not be spouting water vapor. The traditionally mysterious Venus has a dense atmosphere that rains hot sulfuric acid (465ºC). Its thick cloud blanket traps the sun's radiant heat so effectively that the surface of Venus is hot enough to melt lead. That's serious global warming! With this brief look around our solar system we have opened up a bit of hope for primitive life, but found no human-friendly site. This survey is ample reason to be especially protective of life on Earth.

The social structure of Heaven is a fascinating issue. In a darkly sadistic vision one early founder of the Christian Church (Tertullian ca. 155-225 CE) viewed Heaven as a place to revel in the torture of heretics. In imagining more positive happenings, could we expect great food and romance in Heaven? Do we get to pick our age and the ages of our companions? If your long-time tennis partner wants to revert to being 20 and you want to be 50, that difference is a mismatch. It is nice to have the wisdom that comes with age, but if we are revived as elderly, do we also get our teeth back? When a one-year-old child dies on Earth what kind of growing up will she do in Heaven? What of her bypassed youth, the games never played, the friends she never had, the loves never felt? Since Heaven is forever, is it possible to have changes from time to time if situations or relationships don't work out? How do I locate old friends? Do all of us keep getting older, so that no one will be a child except for some new arrivals? I don't know the answers.

How Eager Are You to Reach Heaven?

You can tell something about the character of a person and the sincerity of their religiousness by their attitudes about death and the prospect of Heaven. To what extent does the fear of death lead us to substitute our needs for the facts? Surveys suggest that those with a heightened fear of death are more likely to believe in a literal interpretation of the Bible (Bivens et al., 1994-1995). If anxiety and fear propel us to religious beliefs, it would not be surprising that an increased fear of death might strengthen religious faith in Salvation. Accordingly, after being given exaggerated probabilities of an early death accompanied by somber music and pictures of victims, the religious college students were more likely to believe in an afterlife (Osarchuk and Tatz, 1973).

Ending matters on your own terms. Why isn't self-euthanasia more common among the devout terminally ill? Two prime factors discourage end-of-life suicide. Universally, we are biologically constructed to cling to life. More specifically, neural circuits effective in sustaining life would tend to be passed on to succeeding generations. Your brain has accumulated many such circuits that help you to survive. Our built-in aversion to death is so ingrained it generally overwhelms even a strong faith in an everlastingly pleasant Heaven that might encourage one to make an early exit. Second, Judaism and Christianity both have strong cultural prohibitions against suicide, and our laws, in reflecting public sentiment, also ban suicide. Although the roots of this prohibition are complex, they surely reflect parental intentions to leave offspring who in turn will have their own children. Children and grandchildren are important to most people. At the social level, early suicide hurls relatives into a grieving tailspin and sends a chill through the local community. We may preach the attractiveness of Heaven, but there is no encouragement to show up prematurely. Your time will come. Sensibly from all perspectives, we struggle hard to prevent suicides that leave grief, loneliness, and wrenching guilt.

If the born-again fundamentalists are truly confident they are saved and destined for Heaven, is it their fear of the sin of suicide that prevents them from clamoring to leave this earthly vale of tears and get on with Heaven's eternal pleasantries? Perhaps it is no wonder that few who publicly proclaim their abiding faith in a guaranteed Heavenly afterlife are in a rush to arrive by nightfall.

Even so, in numbers roughly equal to automobile fatalities, more than 100 Americans commit suicide every day. The elderly commit suicide three times more frequently than other age groups. When caught in the throes of a painful terminal illness, it is natural to consider accelerating the inevitable. Consequently, those who believe in an afterlife are more likely to consider passive euthanasia an acceptable option (being allowed a natural death with no heroic life-supporting interventions [Klopfer and Price, 1979]). Following the January, 2006 US Supreme Court decision, the option of assisted suicide continues to be available in the State of Oregon.

Does Our Body Get to Heaven?

It was commonplace in ancient burials that the deceased would be provided with food, drink, jewelry, and various implements useful in the next life. Such rituals raise the question of exactly what survives in life after death. Ascension into Heaven might mean direct bodily ascension, or the ascension only of a disembodied, immaterial soul, or even combining the soul with a miraculously replicated body once in Heaven. Another option is to be reincarnated as another human being or other living creature on Earth. Each of these possibilities raises the question of age, and aging, and choice. Who takes care of infants (embryos?) in Heaven? Or are they immediately made old enough to be independent, say age 18 or 21? If we opt to return to our preteens, do we lose all of the sophistication we have accumulated during our later years? How could you retain your uniquely special personality if you were reincarnated as some other person or creature? Is a person reincarnated as a chicken aware of their prior human existence? If our souls age in Heaven, what experiences do they bring with them and then accumulate? Those who claim they have had contact with departed spirits or have at least seen them, invariably report that such spirits were clothed.

In this short consideration of the afterlife we have unearthed many practical conundrums about bodily life in Heaven. These difficulties suggest we should not think of our bodies in Heaven as a physical reality, but rather consider souls placed in a dream state, with each person's imagination evoking and simulating the pleasures of one's choosing. Because our dreams will be constructed from our own experiences, every Heaven will be personalized. If, like Shakespeare, you had died in the 17th century, your heavenly dream state would have no electric lights, phones, TV, or computers. Of course, you wouldn't miss these pursuits since you had never experienced them on Earth. A dream state would require a brain-like central processing unit that could recall stored memories and embellish these with fantasies. It seems almost arrogant to say that only humans have souls, since we are part of the community of mammals. I myself am convinced my German shepherd dog could think, anticipate the future, feel elation, sadness, and guilt. I would enjoy the company of its soul or imagined soul in another life. The concept of a mobile soul was commonplace in primitive cultures. "[The Mayans] believed that after death there was another better life that the soul enjoyed after it left the body. They said that this future life was divided into good and bad, into suffering and peace. The evil life full of suffering, they said, was for the wicked; the good and pleasurable one for those who had led virtuous lives," (de Landa, 1572, p. 95). Much earlier, Greek philosophers held that humans were the union of a temporary body with an immortal soul. Today, geneticists similarly view the body as a temporary repository of quasi-immortal genes. The gene's way of unintentionally approximating immortality for each of us is to be expressed in the bodies of our future descendants on Earth. Genes aside, perhaps the best that a weightless, formless, eternal soul can look forward to is a nirvana somewhere in the timelessness of space—a vivid dream state, or less appealingly a perpetually mellow state where we are zoned out forever.

10

&

The Evolution of Religions

Religions Began Early and Often

Anthropologists have eagerly sought out primitive tribes as "living fossils" which might reveal basic human conditions uncontaminated by modern civilization. Although fascinating, it remains an uncertain exercise to consider the rituals of today's indigenous peoples as proxies for ancient religious rites in preliterate times (Durkheim, 1915; Frazer, 1915; Johnstone, 1997). By unwritten convention anthropologists no longer discuss the deep origins of religion for fear their reputations would be tainted by the "if I were a horse" error. To guess what drove primitive peoples to religion is like asserting, "If I were a horse, I would want to have green grass and clean water with room to romp in the company of other horses," as if we know what horses might be thinking (Bowie, 2006).

While admitting that the deep origins of religion are speculative and are likely to remain so, I am unfettered by the social norms of anthropologists. So, I find it worthwhile to imagine how primitive humans might have dealt with the realities of life infused as they were with uncertainty. Surely, a bountiful world of plenty that knew no harm, sickness, or death would reduce the incentives for spirituality. Religious beliefs and activities would be quite different had they been driven solely by positive communal benefits, like hunting and gathering, bonding, child rearing, and other social functions. A substantial part of religion represents dissatisfaction with life's unpleasant features. Bronislaw Malinoski (1884-1942) believed that the fear of death was the most basic instinct that contributed to religious mythology. The world of primitive tribes like the Ixtepejano of southern Mexico is saturated with concerns about harmful spirits (Bowie, 2006, p. 8).

Religion is an agency for confronting existential problems. Constantly imperiled by harsh conditions, survival was problematic for early humans. Thomas Hobbes (1588-1679) famously described human lives in the natural state as "solitary, poor, nasty, brutish, and short." Where were survivors to turn in their grief over a premature death from a mysterious sickness?

Early signs of religious activity. The earliest religious sites were probably at graves. The earliest known graves date from Paleolithic times, some more than 100,000 years old. The religious practices 60,000 years ago provisioned corpses with medicinal herbs and weapons ready for an apparent afterlife. Egyptian tombs dating from 2100 BCE provide detailed evidence that a comfortable afterlife was intended. Egyptian royalty who could afford it made elaborate preparations to ensure every indulgence. The royal mummy was surrounded by a variety of accurate wooden models, each about two by two feet square. There was a kitchen, brewery, granary, weaving shop, copper-smithing shop, and other models complete with miniature workers. Magic was expected to transport the shops and their human figures to Heaven where they would expand to full size and spring into action to produce the variety of food and comforts that royal pleasures demanded. An impressive replica of such an Egyptian gravesite may be seen in the Ny Carlsberg Glyptotek Museum in Copenhagen, Denmark.

Early religious ceremonial sites were also commonplace. Outdoor sites with massive markers are relatively recent, like the 5000 BCE circle of heavy horizontal capstones somehow heaved on top of vertical slabs at Stonehenge in rural England. Such imposing ceremonial sites suggest that even before civilizations were capable of written records, the ceaseless longings of humans probably had already created numerous local religions—with their rituals practiced at dedicated sites.

As waves of humans migrated out of Africa and pushed across the plains and mountains of Europe and Asia, their dispersion and seclusion allowed local tribes to put down unique spiritual roots. Such geographic isolation provided dispersed testing grounds for emerging religions. Group cooperation fostered by religion helped to cope with unpredictable threats including attacks by hostile tribes. While many religious groups could claim success in forging unity against marauding tribes, only groups fortunate enough to settle in a benign and fertile region might markedly reduce sickness, disease, or childhood mortality. Although there is no direct evidence, it is reasonable to assume that religious credos arose as soon as language skills allowed explicit communication of personal fears and concerns. At the time when early religions established their doctrines millennia ago, many natural features like fierce gales, thunder and lightning, fire, sickness, and death were awe-inspiring mysteries.

Monotheism was a late arrival. Even at the height of classic Rome, humans continued to prefer multiple gods, each with a specific domain of power. With their differing roles and personalities, these classical gods were inclined to express human-like weaknesses such as rage, jealousy, passionate loves and greed. In squabbling among themselves and cavorting with humans, the multiple gods of ancient Rome undercut the dignity of the Deity and the authority of the Christian Church.

Earlier, in the complex texture of some agricultural settings it may have made better sense to have several lesser gods, each with a specified responsibility for rain, corn, beans, insect pests, productive chickens, fertility, fire, health, war, and so forth. Linguistic records of goddesses of fertility include Aphrodite, Artemis, Astarta, Halthor, Hera, Isis, Tanit, Turna, and Una and more. Perhaps it was easier to accept the limitations of a restricted demigod than to rationalize why one over-arching God would be so fickle as to shift neglect from one responsibility to another—a single God of apparent whimsy. Of course, the single western god was frequently assisted by coveys of angels and several thousand saints keyed to different days and circumstances. The supernatural world tends to expand with time.

Religion today, at least among the educated, is no longer driven by a need to understand the operations of the material world, for these have been largely explained. In clarifying causation in the natural world, science has eliminated or moderated some primal religious superstitions. The progressive demise of speculative demonic forces is evident in the historical shift from grappling with multiple evil spirits to modern religions inclined to consider supernatural spirits as allies rather than adversaries. Disconcertingly, careful modern inquiry has failed to illuminate the supernatural world of spiritual figures and with it the prospects for life after death.

Religious Rituals, Pilgrimages, Doctrines, and Strategies for Expansion

Religions differ in the details of their connection to the supernatural, the literalism of Holy Scripture, the path to Salvation, the rituals and beliefs held in common with mainstream religions, the charisma of the leadership, the commitment to a top-down authoritarian structure, the demands or requirements of membership, the reproductive fertility of its members, and the duty of members to gain adherents through evangelism.

Rituals. At a minimum, prehistoric religious rituals probably involved food, fire, venerated objects, and some sacrificial offerings. The diversity of modern religious rituals reflects the diversity of human spirituality. Some people pray in tranquil solitude. Others find reaffirmation in music: the grandeur of familiar hymns, the joyous feelings of swaying spirituals, or the complex beat of modern rhythms. Still others exalt in the inspiration of incense and incantations beneath towering stained-glass windows in a lofty cathedral. When adherents have communal celebrations, their surging emotionality makes them believe more fervently in the supernatural.

Pilgrimages. Ancient pilgrimages must have begun with small gatherings. When thousands of believers make an exhausting journey to a sacred site, it heightens the intensity of the religious experience. The massive hajj or pilgrimage to Mecca is the most famous. The largest in Spain is the pilgrimage of the Virgin of El Rocío that occurs during Pentecost, seven Sundays after Easter. Its origins date from 1280 CE after the unearthing of a buried statue of the Virgin Mary was soon linked to various divine apparitions. Every year huge crowds of modern pilgrims gather in the tiny village of El Rocío near the National Park of Doñana on Spain's southern Atlantic coast. For nearly a week one million pilgrims arrive by foot, oxcart, horse, or carriage, straggling in through dust and mud and oppressive heat to bed down in the El Rocío outpost which resembles a Hollywood western frontier town complete with hitching posts in front of each hotel. Brimming with round-the-clock parties, celebrants bless some 95 Madonnas brought in from participating churches in a procession that climaxes with the statue of the Virgin of El Rocío, also called the Queen of the Wetlands or the White Dove. As with all pilgrimages, the clerical intent is to provide collective spiritual renewal.

Doctrines. A formal set of religious beliefs is a doctrine. Doctrines ensure the marking of life's major stages: birth (baptism), coming of age (confirmation), marriage, children, and death. The periods between are to be filled with involvement in church affairs, especially on a day of worship. Doctrines are the many threads of religious engagement woven into the fabric of human lives—denominational ties that bind. Uppermost are the details that promise a pleasant afterlife. Success in sustaining the commitment of the membership is the practical key to evaluating the utility of doctrines.

Expansion. An excellent path to enlarging a religion is to fan out and spread the Word much like the travels of Paul and his companions. Gaining converts through proselytizing activity is a major part of the Mormon religion and other sects now experiencing growth. Evangelical sects that corral the most members are likely to win out in the marketplace of religious competition. According to the Catholic Church, Peter inherited Christ's authority and passed it down through a succession of popes. This tight linkage to Christ seems like a distinct advantage over other denominations, yet Catholicism has been losing out to new, faster growing sects. In Latin America the beguiling ways of evangelical and charismatic sects are seducing members away from traditional Protestant and Catholic Churches. Though they customarily meet in modest stick sanctuaries, the upbeat emotionality and inflated promises offer a spectacularly easy route to the top of Mount Salvation. How should the keepers of traditional institutional religions attempt to counter this charisma? The competition for souls and the difficulty of disproving claims about the supernatural combine to create a nearly irresistible temptation for mainstream clergy to offer hyped benefits—alluring promises that no one can verify.

Fierce competition in any field can debase ethical practices. When an intense political campaign heats up, truth often suffers. Religious sects are also vulnerable to exaggeration, even inventing claims to gain an advantage. For politicians and religions to win—to gain the support and votes of adherents—they must offer a more appealing image than the competition. The wings of Christianity now soaring in ascendant popularity are the Fundamentalist, Pentecostal, and Charismatic churches. Break out in joyous song in the solidarity of other born-again souls. Entry into Heaven is essentially free; you need only be born-again in accepting Jesus as your personal Savior. Faith alone will do; weekly confessionals and good works are not required. One can shed the stigma and burden of original sin and committed sins, so prevalent in Catholic and Calvinist traditions. Religious competition can lead down a path that bypasses complexity, if the traffic will go there.

Genes for Religious Belief?

Is the widespread prevalence of religious beliefs and practices evidence that religiosity has conferred some selective advantage? This argument would gain more force if it were possible to identify "religious genes" linked to spirituality-inducing hormones or neural circuits. The evidence remains thin that human brains are hardwired for religion (Azari et al., 2001). Certainly, not all human brains are predisposed to feelings of deep religious experiences (Armstrong, 1993). If one type of serotonin receptor in the brain is the product of a "God gene" as Hamer (2004) labels it, its actions are certainly not robust, for it accounts for only 5% of the variance in human "religious" behavior.

The Success and Failure of Promulgating Religious Doctrines

Religion is today's most immense human enterprise. Consider the scope of religious membership, financial support, and belief. Of the 6.5 billion humans on Earth there are 1.2 billion Muslims, 2 billion Christians, 1 billion Hindus, 0.4 billion Buddhists, and 0.014 billion Jews. The Christian community operates on a collective annual budget of 28 billion dollars aided by 10 million workers. These dedicated workers have been effective in convincing the vast majority of Americans that a pleasant afterlife awaits them in Heaven. Over half of Americans also believe Hell exists, though only a few believe they personally will end up there. (Don't ask these few what they have been up to!) Religion takes on enormous importance in many lives (Barrett et al., 2001). Tenacity is one of the most impressive traits of successful organisms and religions. The fittest have survived and flourished—the great majority has perished. Most religious sects probably faltered soon after they began—lofty aspirations that fell and rapidly expired due to marginal leadership and thin doctrines.

Today, an aspiring religion may begin as a personality cult that withers under informed scrutiny. Thousands of religions have begun only to languish after failing to attract enough members (Stark and Bainbridge, 1986). Even so, any religious practice that manages to enhance human procreation and survival, even inadvertently, has some potential for sustained growth. The long-lasting institutional religions are enduring outcomes of natural selection. The successful denominations have evolved doctrines and techniques that assemble believers—generation after generation. Outside critics may view such diverse doctrines as persuasively crafted theological errors; misinformation used to provide the powerful with control over the gullible.

The remarkable persistence of religious commitment leaves virtually no irreligious communities on Earth, with the possible exception of countries where collective religious observances are suppressed by government edict. For example, in Communist China today few people are familiar with Confucianism. Even though the Soviets forbade the practice of Christianity, it managed to revive, but not thrive, after the fall of communism in Russia and the former Soviet block in Eastern Europe. A decrease in religiosity, especially in younger generations, appears to be linked with Europe's increased standard of living as measured by longevity, education, and per capita income (Norris and Inglehart, 2004). The high standard of living in Japan's eastern culture may also have reduced religious behavior. Fewer than 10% of Japanese believe in God. Japanese often concentrate on venerating the graves of their ancestors as a way of connecting with those who have passed on. Although an active cadre of Buddhist and Shinto priests maintains thousands of temples and shrines, tourists are often the most obvious visitors. Yet, even today the Imperial Emperors of Japan rule by divine right as descendants of the Sun-goddess, Amaterasu.

Many of today's mega-religions can trace their origins to a cult that managed to prosper. "Cults are groups that center about charismatic leaders who work to see that their members are separated from society in general, from family members, and from anyone who might offer divergent views" (Spilka et al., 2003). Jesus explicitly asked his followers to separate from the unsupportive and the unpersuaded. In most instances cults fizzle out and the surviving members disperse to gather up the remnants of their lives. In the worst instances members perish in mass tragedies like the Jonestown People's Temple and the Branch Davidians. Fortunately for the local community and for the future growth of Christianity, Jesus decided to sacrifice only himself and not his disciples, who survived for a time to spread the good Word.

Every new religion has a high risk of failure. Christianity clearly struggled to survive after the death of Jesus. The local Jewish populace remained more satisfied with their well-established traditions under Moses. While Paul's traveling ministry was making Christianity attractive to Gentiles around the Mediterranean, the bishops were busily honing Christian doctrine to enhance their power and gain adherents. Adroitly promoted religions fill the niches offered by human existential needs. A religion must appeal to peoples' needs and longings or it will fail. Those doctrines that worked to strengthen the tribe and enhance human survival were more likely to be passed on to future generations. Successful sects benefited from astute leadership in addressing human needs, and from luck in avoiding the toll of wars and other disruptive events. As communication improved, local religions expanded through takeovers and mergers with neighboring denominations.

The three major western religions (Christianity, Judaism, and Islam) were founded in pre-scientific times when active supernatural forces were widely accepted. Indeed, evidence for the supernatural was expected as part of a credible showing by religious figures. In order for a religion to gain acceptance it was necessary to have a strong leader who offered captivating, frequently miraculous, displays of supernatural interventions. The evidence for the supernatural took many forms. Divinely inspired writings or tablets inscribed by God himself, accurate prophecies by inspired prophets, as well as miracles, visions, and the answering of supplications and prayers—all of these implied on-going communication with God. In biblical times miraculous healings, prophecies, and revelations were the staple of many itinerant preachers. Even so, skeptical onlookers were ever alert for possible deception by charlatans or satanic agents.

As they attempt to garner support and create inspiration, today's religious leaders offer ancient narratives enlivened with recent tantalizing evidence of ongoing supernatural actions. Present-day claims of ongoing divine influence must meet higher standards because today's citizens are better educated and more informed about causation. As scientific facts have accumulated and procedures for testing evidence have advanced, religious doctrinal claims have become less sweeping. The paucity of credible supernatural interventions, the accumulating explanations of science, and the retreats of outmoded religious dogma are an unholy trinity for many of the faithful. Science generally prevails in material matters where the facts and conclusions are open to evaluation and replication by others and their correctness can be demonstrated both in accurate predictions and in further applications. Theologians strive to frame an appealing set of beliefs that will remain out of the reach of science. Souls have no mass or energy, although they are capable of entering bodies undetected and can ascend or descend at the end of life while retaining full capacity for thinking and feeling. Original sin is invisible.

Generally there is a tradeoff between the ability to verify a religious assertion and its spiritual significance. Credible religious claims are usually modest claims. Credibility sags when extraordinary claims are raised, like claiming that earthly intercessory prayers to God reliably heal serious disease, or that a particular individual can perform miracles, or that one can have conversations with deceased relatives or even a chat with God. Church officials are well aware that future scientific advances may deflate carelessly launched doctrinal claims. The late Pope John Paul II maintained a science watch unit to insure that the Catholic Church would not be blind-sided by new scientific discoveries. This Pontifical Academy of Sciences is an advisory council of 80 members that includes many distinguished scientists. The preservation of institutional religion is a serious business.

Early Preachers and Their Success

In establishing spiritual authority, charisma mattered more than credentials. Dating at least from the time of Moses, every major religious leader has claimed to communicate with God. Christ was called the "Son of God" (*John* 3:18) and "God" (*John* 10:30) among dozens of similar divine references. It was the angel Gabriel who told Mohammed the oral narratives later set down in the sacred text of Muslims, the Qur'an. Joseph Smith, the founder of the Mormon religion (Church of Jesus Christ of Latter-day Saints), had a vision as a teenager that he was a prophet of God. Communication with the divine typically continued throughout a prophet's career.

Box 10.1 Brain Disorders and Religion.

It is not unusual for lone preachers to be demanding and narcissistic autocrats with a suspicious, paranoid outlook. Such a warped personality can be quite functional in its context. Some religious careers have been propelled by frank mental illness or neurological dysfunction like epilepsy. The public can be spellbound by behavior driven by hallucinations. Ellen White was hit in the head by a hard snowball and thereafter reported visions from God on behalf of the Seventh Day Adventists she founded. An aura or strange vision followed by fainting and convulsions due to an epileptic seizure has been taken as a sign of the presence of the Holy Spirit, or a satanic spirit, depending upon the circumstances. Now that more is known about disordered brain function, bizarre behavior is less likely to mark an individual as possessed of God. For example, auditory hallucinations of a religious nature can result from a person's own sub-vocal speech, incorrectly attributed to an external voice (Bentall, 1990). In Salem, Massachusetts the disordered thoughts of schizophrenic women indicated possession by the devil whose expulsion required torture or burning at the stake. Even today, some individuals find evidence for possession by satanic spirits in certain mental confusions and obsessions. This can lead Catholics and others to perform exorcism rituals.

A messianic leader offering relief from earthly sufferings while compellingly pointing the way to eternal bliss represents a new religion's most important asset. Such charismatic leaders must be absolutely self-assured. They dispense healings and miracles and other evidence of direct communion with God in return for requiring unswerving commitment from adherents. The most effective preachers are extraordinarily good at these tasks. They rely upon several crucial techniques. Establishing a culture of absolute belief is the key. An entrepreneurial preacher must also be captivating enough to gather and energize groups of followers who will act as disciples and work to spread the faith. Sermons ought to contain wisdom in woolly parables to ensure dependency upon the preacher for clarifying interpretations. Through reward and punishment preachers tap into many motivations: hope, despair, anxiety, fear, guilt, love, comfort, and solace.

As Jesus found, it is helpful to disparage competing beliefs and loyalties. With the now classic advice, later dispensed by those recruiting for monasteries and nunneries, Jesus instructed followers to ignore their family, quit their jobs, give all worldly possessions to the poor (via the church), and then bond with him. In addition to requiring a serious commitment, successful prophets motivated their flock with credible evidence of power through prophesy, healings, and other miracles while stirring up strong feelings, swollen by the group setting. Successful religions capture the heart. Imagine how little appeal Jesus would have had as a quietly unassuming teacher shorn of miracles. Beyond his wise advice, which still informs Christians, he needed to be able to make accurate prophecies and perform miracles that would excite. After his death church leaders felt it necessary to demonstrate Jesus' strong connections with the supernatural. Perhaps the most important elements for the success of Christianity were declarations of his divinity. Jesus did not claim to be God's son in any literal sense. But the narratives grew that he had been born of a virgin (it was not Joseph, but God as the Holy Spirit, who somehow deftly donated at least a Y chromosome to Mary's egg), resurrected after death-by-crucifixion, and subsequently disappearing—apparently ascending into Heaven.

Miraculous events might seem to be inherently persuasive, but with Satan's sorcerers also out-and-about performing their own dark miracles, it was no easy matter in the holy land to prove that a miracle had divine origins. Satan and his allies, his dark angels, were apparently filled with formidable power. During the decades surrounding biblical times numerous preachers claimed to perform miracles similar to those of Jesus. Many were false prophets, even fakers, who only pretended to perform miracles. But others in addition to Jesus may have been authentic. As the *Acts of the Apostles* of the New Testament relates, Peter raised Dorcas from the dead. Paul revived a spectator from a fatal fall. Paul also healed the diseased, restored the lame, and eliminated palsy (*Acts* 8:7; 14:10). However, these powers could be harshly applied, for in at least one instance, rather than heal, Paul intentionally blinded a miscreant as he uttered these words, "And now behold the hand of the Lord is upon thee, and thou shalt be blind" (*Acts* 13:6-12). In addition to miracles, another crucial talent for a successful ministry was skilled prophesy that persuasively implied supernatural powers. Some preachers claimed they themselves were the very embodiment of Old Testament prophecies made years before—the long awaited Messiah had arrived. The public generally met this claim with considerable skepticism.

Preachers and prophets who impart uncertainty will not succeed because the psychological needs of a suffering populace require persuasive reassurance secure in purpose and belief. The aspiring prophet who admits to the many uncertainties about life and death, Heaven and Earth may soon be in need of some other line of work. Converts are won through the certainties of exhilarating revelations and the received Word from virtually direct communication with God. The long-established religions grounded their doctrines in divine personages and miracles from ancient times—claims that will never be available for modern assessments. Imagine if someone today insisted he was the Second Coming of the Lord Jesus Christ. Confirmation would be quickly sought in a DNA sample. Such a joyous return could provide a stunning opportunity to examine the presumably flawless chromosomes of God.

Only the most skilled prophets can get away with consistently hyped claims. Even so, the temptation for religious trickery is always present. Sophisticated manipulation is with us today bundled into cleverly refined and extravagant presentations. Fraudulent faith healers typically use some ruse like a confederate secretly radioing crucial information picked up by a tiny receiver hidden in the ear of the "prophet." Billy Joe Harvis, Jimmy Swaggart, Robert Tilton, Jim and Tammy Faye Bakker (Pat Robertson's choices for the first hosts of the "700 Club"), and Ted Haggard, ex-leader of 30 million American evangelicals are notable Christian figures who have throbbed with chicanery. Is it ethical that religious leaders like Pat Robertson and the late Jerry Falwell have repeatedly followed a calculated pattern of drumming up publicity with outrageously harsh Old Testament accusations, soon to be followed by feigned apologies?

The Mass Media Provides Preachers With a Wider Audience

Gutenberg could not have imagined how his printing presses (1458 CE) would open up floodgates of religious information. Considering that it might take a monk 20 years to make one attractive hand-written transcription of the Bible, the ability to rapidly print multiple copies of the Bible dramatically expanded its distribution. Gutenberg's presses certainly aided Luther's cause in distributing his 95 theses. Today, more than ever, to pour through religious tracts—affirmations of faith, church encyclopedias, accounts of saintly lives, religious instructional books for all levels, and rampant commercialization of religious propaganda—is to be inundated by torrents of advice intended to persuade and slake our spiritual thirst. It would require the entire shelf space of more than 100 typical high school libraries just to display one copy of every religious book title now in print, most of them pressing their doctrine with relentless zeal. Given the surge of new religious denominations and avid readers for millions of religious books, it is evident that religion remains an animating passion, especially in America.

Regardless of size, small denominations, minor sects, one-of-a-kind starter cults, even mainstream religions, all share a common concern about expanding their membership. Events calendars enriched with entertainment, and even TV spectaculars, are some tempting approaches. The telecasting of huge Sunday services is especially captivating for those who emotionally confuse the display of earthly opulence with Godly power. For instance, the attractiveness of social amenities is evident in the Willow Creek Community Church in South Barrington, Illinois. At the entrance door to Pastor Bill Hybels' office is a sign with three questions borrowed from Peter Drucker, the famous management guru for business and charitable organizations. "What is our business? Who is our customer? What does the customer consider value?" By emphasizing activities and facilities rather than strict religious doctrines and symbols, this mega-church has more than 17,000 members. Robert Schuller's huge 20-million-dollar made-for-TV worship facility has a glass roof and glass walls ideal for television viewing. An ostentatious TV church service decked out in wealth, and a mega-church with huge numbers of participants both invoke the message that "might is right."

Institutional Religion, Self-Interest, and the State

A role model? Those hoping to elevate their own denomination to star status should examine the techniques of the Catholic Church. Is there any other institution that has survived for 2,000 years? As a nation state with a diplomatic corps, the Vatican is not a democracy but an authoritarian hierarchical system that has sought to gain a controlling influence in many countries. It maintains vast land and property holdings. The Vatican or Holy See charts the path to Catholic supremacy, and the authoritarian structure of the Catholic Church strives to implement that vision. The Vatican has a vigorous evangelical foreign policy of spiritual conquest backed by billions of dollars of resources and a membership 1.1 billion strong. It is the supremacy of the soul over the body that makes it acceptable, even righteous to concentrate on strengthening the institutional Catholic Church, sometimes at the expense of our present lives on Earth. The major hardships and inequities of secular life such as poverty, the marginalization and exploitation of women, malnutrition and starvation, AIDS and other sexually transmitted diseases evoke some institutional concern. But such practical matters on Earth are secondary to insuring the welfare of the eternal soul in the afterlife. It remains to be seen how much traction neoconservative Anglican theology can gain for the different view that God, relying on the concerted efforts of humans, will rescue Earth and turn it into Heaven (Wright, 2006).

Death by Natural Law? The Catholic Church has viewed it as more valuable to support unfettered human reproduction as a claimed canon of natural law than to maximize the health of infants. Hence, the Vatican has vigorously opposed the use of condoms to prevent infection, including HIV infections (human immunodeficiency virus). As a consequence, HIV-positive males may infect women, who in turn pass on HIV to as many as half of their infants. Maternal transmission of HIV to the child can occur during pregnancy, during delivery, or during breastfeeding. In Africa 1,600 HIV-infected children are born every day or 600,000 per year. Before these sickly children are three years old nearly all will be dead of AIDS, dead of related opportunistic diseases, or dead by abandonment. Largely as a result of the death of the parents to AIDS, by 2015 in sub-Saharan Africa there will be 44 million orphans under the age of 15, about one-third of whom will be HIV positive (Stine, 2003). Some Catholic clerics, including two cardinals and the Catholic bishop of France in 1996 urged the use of condoms as the lesser of two evils. Their concerns did not persuade Pope John Paul II. In the spring of 2006 the news media began reporting that liberal segments of the Catholic Church were urging Pope Benedict XVI to allow condom use for heterosexual married couples. Regardless of the Vatican's stance, it is fair to say that AIDS would have become a nearly unprecedented public health dilemma. (Malaria and hepatitis C are comparably prevalent.) In the 20th century the solemn obligation to defend and enhance institutional Catholicism sometimes triumphed over humane objectives. Claims of working for the grand triumph of the eternal soul will always trump ephemeral misfortunes on Earth. Although many Catholics believe its church has slid far off the moral path in these matters, Catholicism's persistence for two millennia suggests that Vatican policies have been well-adjusted for its survival.

Box 10.2 Frankly Facing the Tragedy of AIDS.

In a quarter of a century the AIDS pandemic has already infected over 90 million people and killed more than 30 million. By 2010 in AIDS-impacted Botswana in southern Africa the average person will die before age 30. The death of both parents has left millions of orphans, many sick and dying of AIDS. The human side is evident in two personal stories. First, Alicia, a mother dying of AIDS. Alicia is knitting a keepsake remembrance scarf for her two-year-old daughter Maria, unaware that Maria is also infected and will die even before she does. Elsewhere in a small village two boys, 10 and 12 years old, take care of their AIDS-infected mother. They make breakfast for the three of them, get themselves to school, and return in time to make dinner for their bed-ridden mother. After giving their mother a bath they use soft towels to gently dry the oozing sores on her skin. Then they carry her back to bed and get set for tomorrow's effort at schooling and survival. These sad lives reveal the truly innocent child victims of the AIDS pandemic (After Stine, 2003).

"An Ode to AIDS"
One breath at a time my child, I will stay alive
Yet, somehow you must sense the death in my eyes
Mommy, will you always come back?
If you die mommy, where will I go?
Mommy, why do they take my blood too?
Do children die who are only four?
Oh mommy, stop retching on the floor.

State and religion: authoritarian titans. It is ironic that powerful attempts to silence dissent often end up by strengthening belief. In 1873, Otto Bismark carried out a culture war (Kulturkampf) against the Catholic Church. He expelled the Jesuits from Germany, closed religious orders, and placed the state in charge of Catholic schools. One notable successor foresaw this as a naïve action with little lasting effect. "For the political leader the religious doctrines and institutions of his people must always remain inviolable; or else he has no right to be in politics, but should become a reformer, if he has what it takes! Especially in Germany any other attitude would lead to a catastrophe" (Adolf Hitler, 1925, p. 116). As we will see this may have been an early attempt by Hitler to gain popularity while hiding his negative attitudes toward religion.

It is risky for religious institutions to oppose the devaluation of life by authoritarian governments. Even a brief look at the tussle between Nazism and Christianity is humbling. Because the Catholic Church had been frequently criticized for its tepid opposition to Nazism, Pope John Paul II offered some apologies in 1992. In the run-up to World War II the Vatican acquiesced to the broader objectives of the Nazi Holocaust which included the extermination not only of Jews but also of gypsies, homosexuals, and various malformed, mentally ill, and mentally retarded individuals. For example, the Catholic Church agreed to release its genealogical records to the Nazis. This made it much easier for the Nazis to identify Jewish heritage in selecting individuals to send to concentration camps. One would think that after World War II, it would have been an appropriate response to excommunicate Catholic Nazis bearing high-level responsibility for the evils of the Holocaust. Instead, Vatican officials helped Martin Borman,

Adolf Eichman, Franz Stangl, and many less notorious Nazis to escape to secret hiding places, notably Catholic South America. As of 1975 not one such Nazi had been excommunicated for Holocaust activities, in contrast to the excommunication of multiple religious heretics before, during, and after World War II.

To be fair, not all Catholics were compliant with Nazi wishes. In the ramping up of Hitler's power in the decade before World War II, some Catholic officials in Mainz, Cologne, and Bavaria openly opposed the Nazis, calling its racism "a grave error" and its hyper-nationalism "a religious delusion." As Hitler gained power, he exercised it through arrests, searching the homes of priests, confiscating property, and disbanding dozens of Catholic political societies. The intimidated Catholic officials largely ceased their opposition. On March 28, 1933 in their concordat with the Nazis, the Catholic Church agreed to disband a host of politically active groups. Protestant churches were also generally supportive of the Nazi regime. After meeting with Hitler in January, 1934 twelve evangelical officials stated, "the leaders of the German Evangelical Church unanimously affirm their unconditional loyalty to the Third Reich and its leader." Perhaps this had a measure of opportunistic political gloss, but the Thuringian German Christians were surely enthusiastic supporters. They went so far as to characterize Hitler as "God-sent," befitting their motto "The Swastika on our breasts, the cross in our hearts." Beyond being intimidated by the Nazis, and concerned about Godless Communism, both the Catholic and Protestant churches feared losing their substantial public tax support from the German state if the Weimar Republic's provisions for separation of church and state were ever enforced. As state supported religions the Catholic Church and the Evangelical Church received subsidies of 130 million marks in 1933, 500 million marks in 1938, rising to more than 1,000 million marks during World War II. (Likewise, the innovative faith-based government initiatives in the United States seem to have purchased the political allegiance of some religious sects [Phillips, 2006].) Obviously, official Catholic or Protestant opposition to Hitler would have come at a huge financial cost. Already by June of 1934, Hitler's intentions had become more evident when his goons murdered Dr. Fritz Gerlich, editor of a Munich Catholic weekly newspaper and Dr. Erich Lausener, General-Secretary of Catholic Action, and several others who were deemed "anti-German." Hitler's subsequent actions also failed to support his early claims of being a Christian. In private remarks Hitler said, "One is either a Christian or a German. You can't be both," as he voiced his intention to stamp out Christianity "root and branch" (Johnson, 1976).

Looking back, it is easy for us now to believe these Christian church leaders were moral cowards who sold their loyalty. But let's highlight the realities they faced. Consider the plight of the Jehovah's Witnesses who were clearly the bravest sect in openly declaring the Nazi state totally evil. More than 95% were subsequently persecuted and one-third were killed outright for refusing to submit to censorship and to the authority of the national government. It was no surprise that the Nazis ultimately killed the Lutheran Dietrich Bonhoeffer for actively plotting to remove Hitler.

If you are still feeling a sense of moral superiority over the weak-kneed responses of most Christian churches in Nazi Germany in the 1930s, get ready to take a personal test. American citizens needn't go looking elsewhere for oppression given our own gathering dark clouds of intimidation. The present United States administration practices warrantless invasions of privacy, secret jailings and disappearances, renditions and torture, and other types of totalitarianism. To test your superior courage and righteousness I suggest you wear a shirt that says, "Whom Would Jesus Torture?" and see what happens when you try to board a U.S. domestic plane flight. Well, why are you hesitating to exercise your constitutional right to exhibit a Christian message? Is your courage sagging over even this modest display of moral principles? Could we have counted on you had you lived in Germany in the 1930s? Authoritarianism is intimidating, wherever it occurs, whether the governmental leader is avowedly secular or makes pious noises to mask evil.

The Three-part Strategy of Successful Religions: Conquest, Message, and Faith

Conquering the competition. The durability of particular religions reflects the survival of the fittest. A durable religious organization should be armed with superior doctrines and management to out-persuade the competition. In its earliest days Christianity had to compete against well-rooted paganism and its glorious marble temples erected to honor gods like Apollo and Athena. Consequently, when Christians gained power, they constructed churches by razing and replacing splendid ancient temples standing in homage to classical gods. Centuries later Christianity also smothered pagan temples in the Americas. "Covered in thick forest, we cleared this [off the Mayan temple], and on it, using its own stones, we built a proper monastery all of stone; and a fine church we named Mother of God," proudly recounted Bishop de Landa (1572, p. 142). The strategy of overtopping pagan construction served two purposes. Demolition of a pagan temple obliterated reminders of the competing religion and also freed up finely hewn limestone or polished marble for the construction of a Christian church on the same site. Even when ancient monuments were allowed to stand, their statuary and friezes were frequently defaced to eliminate reminders of pagan beliefs. You may also recall the similar strategy of Afghanistan's austere Taliban Muslims who recently destroyed ancient Buddhist statues they viewed as offensive competing icons. One of the most evocative in Bamiyan, western Afghanistan was the world's tallest at 175 feet high.

On the educational front competing religions may target the same youth groups for indoctrination. In the arena of religious competition for political governance, it is evident that continuing tensions could expand into large holy wars, particularly where land and water are limiting. Disputes over religious doctrines can magnify tribal hatreds.

Proclaiming a persuasive message. As towering prophets, Moses, Jesus, and Mohammed launched the three dominant western religions. The Old and New Testaments encourage elaborate procedures to identify other credible prophets. In Catholicism accurate prophecies were considered a particularly powerful sign of divine connection, especially if the prophet had been closely scrutinized and certified by the Apostolic Church. A true prophet ought to be a trusted member of the institutional church who displays signs of being filled with the Holy Spirit. No matter how accurate they might be, revelations from self-appointed prophets do not qualify. The actual prophetic claims, as prompted by a vision or other revelation of the candidate prophet, must be witnessed by two or three credible individuals and in turn must be interpreted by an existing prophet of the Catholic Church. An additional church prophet must judge the event and certify it was a successful prophecy. This layering of validation by church figures cuts down on accidents or the ambitious intentions of false prophets.

After church officials had confirmed that the prophet's credentials were in order and God was the source of information, it was nonetheless acceptable for aspects of the prophecy or prediction to fail later. Why? Because, you see, God might have changed matters in the interim between the prediction and the prophesied event. Equally importantly, it was wholly unacceptable to make a true prophecy while teaching a false doctrine. You can imagine the threat to an established religion from an individual empowered by accurate prophecy who proceeds to elevate a competing religious doctrine. Death was the traditional penalty for this crime under the laws of Moses. In more recent times the punishment has become dis-fellowship from the Church, which is bad enough, since one is sent to Hell for an eternity with no hope of parole. In summary, strict rules for validating modern prophecies allow the church to enhance its standing while preventing the infiltration of doctrinal error or a budding schism engineered by independently ambitious individuals.

Generating an overriding faith. To have faith requires that you adopt a stable doctrine. It is most effective to indoctrinate the child before normal skepticism sets in. From the perspective of the indoctrinators, there is no better time to install religion than the age of youthful innocence. To fill youthful vessels, precious small children, requires a carefully crafted positive message, a representative of the church as a conduit to God, and perhaps the added motivation of fear of damnation. Parents and clergy are often united in training naïve children in the proper religious practices and beliefs. The warm memories of childhood religious experiences can help a grown adult sit though a set-piece service held amidst faded icons in a dank cathedral.

Debate continues over the proper limits of indoctrinating a child into the fundamental beliefs of a specific religion. Is it child abuse if the once-indoctrinated child becomes traumatized when disbelief later sets in? Such unraveling of religious dogma can be wrenching. But isn't each family entitled to teach its children about the superiority of its own religion? If early religious education were to praise multiple religious faiths, the child might fail to hold their parent's religion in special regard. Perhaps the key word in this discussion is "vulnerability." Young children are innocent and vulnerable; in normal upbringing most young children follow parental dictates with little skepticism. Children who trust and obey their parents have a better chance of survival. Indifferent parents may put children at risk. Everyone agrees that parents must provide some socialization of the child or risk molding a dysfunctional adult. Certainly, the training of some ethical values is important. In reality, a particular child will almost surely receive religious indoctrination if its parents are absolutely convinced by faith in the truth of their vision of religion. But narrow and intense religious indoctrination may blinker the child's future capacity to deal with emerging understanding.

For long run success religions must offer messages consumers want. It is helpful to provide evidence of modern revelations or even miracles to reassure church members that even "today this church, your church, is actively embraced by God." Prayers successfully answered, miraculous escapes from disaster, religious statues that stood proud through a fierce storm, recoveries from illness, family successes—all of these are cited as evidence for the presence of a personal God.

One can be awed, even transformed, by a velvet-robed ecclesiastical figure climbing a spiral marble staircase in an immense gilded Cathedral. These displays of grandeur and power provide a sense of a mighty fortress overseen by a supernatural power. Perhaps such settings add to the wide popularity of regal religious services crafted especially for television. A quite different successful strategy is to have a casually dressed minister talk in a youthful jargon to engage adolescents. Religious promotional efforts are sophisticated and extensive. Billions of dollars are spent marketing religion for all ages through the mass media, including magazines and newspapers, radio, TV, and the Internet (www.academicinfo.net/religindex.html) and in millions of religious books, including eight billion copies of diverse versions of the Bible. Virtually every hotel room in America has a Gideon Bible.

The developed world has an abundance of material comforts, significant reductions in disease, effective treatments of common acute illnesses, and generally peaceful, law-abiding societies. These advances over primitive times have reduced many of our primal fears. Even so, the realities of modern life and death still elicit anxieties. A core aim of religion is to convert such feelings of uncertainty into secure beliefs and close spiritual harmony with the church. In the United States, where skeptics are in a minority, faith is a popular means for quelling anxiety, especially in certain geographic belts. Church officials know that when the needs are great, dogmatic faith can override clear and tangible evidence to the contrary—facts are no match for denial driven by anxious concerns. Institutional religion, ambitious preachers, and politicians are aware that the tension associated with pre-marital sex, abortion, homosexuality, and same-sex marriages are useful tools for gaining religious or political support. In many nations religious practices and beliefs are mandated by state sponsored religion—faith by fiat. Likewise, as described in Chapter 18, in a virtual retreat to medieval times, some fundamentalists want the United States to become a Christian theocracy (Phillips, 2006).

Longstanding Religious Doctrines Meet Human Needs: Changes in Limbo

Intermediate stops on the way to Heaven. Church practices evolve in response to the evaluative question, "Overall does this particular practice enhance the standing of the church?" In their heyday, the alternative fates of limbo and purgatory provided some flexibility beyond the stark options of Heaven or Hell. For example, an unbaptized child that died would not go to Heaven but would enter limbo, a state St Thomas Aquinas called "natural happiness."

The currently anticipated jettisoning of limbo by the Roman Catholic International Theological Commission is a real-life example of the natural selection of religious practices. Limbo provided a way for medieval Roman Catholics to stress the importance of accepting the church through prompt baptism that cleanses a child of original sin. As Pope Pius X put it in 1905, "Children who die without baptism go into limbo where they do not enjoy God, but they do not suffer either." Today's less intimidated, more questioning populace, has the opportunity to select from many denominational options. Since limbo is conceded to lack biblical backing, it is easy to view limbo as an artificial device that led to indifferent treatment of innocent, unbaptized children. As it gropes with the concept of limbo and the consequences of its likely abandonment, we can hope the International Theological Commission will somehow identify a generous fate for the enormous population of babies still stuck in limbo. Given the high natural rate of early miscarriage, untold millions of human embryos have unobtrusively accumulated in limbo, according to Catholic theology's construction of developmental biology. Not well-adapted to modern times, limbo is no longer pulling its weight for the Catholic Church. Being stuck in limbo certainly does not play well in undeveloped countries having high rates of mortality of unbaptized infants. Nor does it stack up well with the competition from evangelicals who offer more inclusive opportunities to be saved by the power of Christ's redemption. The concept of limbo seems to have played itself out, a loser in the evolution of religious practices.

11

Science, Society, and Religion

Human Fears of the Unknown

There are hundreds of isolated tribes in the highlands of New Guinea, each living in its own lush valley enclosed by steep, rugged ridges. To cross over into an adjoining valley, one must make an arduous ascent, a several day struggle through thick jungle up to the ridgeline. But the New Guineans almost never venture into the next valley because they are quite confident their neighbors would treat them as threatening invaders who ought to be killed. Feeling superior, each tribe is primed to repel any invasion by brutish neighbors who talk in apparent gibberish. In their virtually complete geographic isolation each tribe speaks its own unique language, so remarkably different from its neighbor as to make French and German seem like dialects of one language. In this situation there is a natural fear of an unintelligible band of strangers bristling with unknown intentions. New Guinean tribes understandably feared white explorers with their trucks, radios, and guns—such noisy, frightening mysteries.

To most citizens, science is a mysterious guild that applies incomprehensible methods that produce difficult-to-grasp results. For people of any culture the new and unknown is mysterious and sometimes scary. Even in highly industrialized countries, ordinary folks may fear technology they do not understand. Some fear fluoridated water (you won't find it in Los Gatos, California); while others fear vaccines or even vitamin pills even though these factors were extracted from ordinary plants. Ideally, all of us would all have broad scientific literacy. More realistically, better public understanding of the general process of discovery would at least make science less mysterious and concerning.

Understanding the Elements of the Scientific Process

Uncertainty and credible discovery. Scientists must deal with the uncertain character of naturally occurring events. To cope with some uncertainty, many predictions are given a statistical margin of error as in; "there is a 70% chance of rain today." Random events are wholly unpredictable. Coin flipping is a familiar example of random outcomes. The chemical interactions of ions or molecules can often be viewed as the sum of many tiny random events. However, most large-scale biological events are not truly random in the mathematical or statistical sense. In many situations some outcomes are more likely than others, given an advantage in size or weight or some other factor. "Chance" is a useful term when there is some predictability in an otherwise random outcome. By implying a complete absence of predictability, the term "random" is inappropriate for most instances of natural selection and many forms of genetic change.

The strategy for the scientific exploration of nature follows a short list of proven conventions. Observations and experiments should be carried out with clear and repeatable methods under set conditions to prevent unwanted variations like swings in temperature. This means that control groups occupy a prominent place in the design of scientific experiments. One type of control group receives all treatments and procedures with the exception of the specific drug or other factor being evaluated. If the control group improves as much as or more than the treated group, then perhaps the treatment provided no added benefit—the improvement was a result of chance variation. However, if the size of the treatment effect is too large to be attributed to such chance variation, then it may be a real effect that needs to be considered. Specifically, in an experimental design that uses "weak inference" one can conclude that results are at least consistent with a favored explanation, for example, the cookie on the counter disappeared last night, consistent with hungry mice on the loose. More telling proof comes from designs that use "strong inference" that actually rule out a previously plausible explanation, for example, we can exclude mice as culprits because the cookie disappeared from a jar that had its lid tightly screwed on!

Once the scientific results have been obtained, the data evaluated, and conclusions drawn, it is time for a preliminary airing of the study at a scientific meeting. At large scientific meetings as many as half of the thousands of studies presented may never be properly published. Of course, the better studies are soon submitted to scientific journals. For the most part, work is published only if it is technically accurate and the findings represent a significant advance in understanding, as judged by two or three anonymous experts in the particular field. This so-called peer-review is a significant filter against work that is sub-par. Depending on the journal and the number of studies it receives, the journal's editor will reject 20% to 80% of submissions for various reasons that are mostly related to the quality of the work (Science magazine, 2005, p. 1974). Lastly, when someone repeats the research findings or verifies its conclusions or predictions, it suggests that the original findings and analysis were on track. The standards in science are high—marginal work is likely to be filtered out sooner rather than later. Sound conclusions should lead to predictions that pan out.

The search for understanding in science may begin with a hunch or guess that is then framed as a hypothesis that can be tested by observation or experimentation. Findings that can be repeated or verified are accepted as new facts whose implications depend upon the context. A collection of related facts can yield a generalization or law. In turn, sets of generalizations and laws form the framework for a theory. A theory is a well-tested explanation that unifies a broad range of facts. A towering synthesis of thousands of observations may be required to generate a theory that provides an indispensable framework guiding future research and interpretations. Major syntheses qualifying as theories include quantum theory in physics, atomic theory in chemistry, and evolutionary theory in biology. In contrast to the colloquial meaning of theory as a mere hunch or guess, a scientific theory is a set of general rules that organizes large assemblages of facts. All of us, scientists included, sometimes use "theory" in a more casual sense, like "here is my theory about that" meaning "let me suggest a possible explanation for that." The theory of evolution is not such a personal hunch; it is a theory in the sense of being the topmost framework that pulls together and makes sense of many well-established factual observations and rules.

Scientific conclusions span a wide range of credibility, from the well-accepted and well-established, to newly published discoveries awaiting confirmation by others. Even well-supported scientific facts may be superseded, as new findings provide better-informed or more precise understanding. The assessments by experimental science are open to revision.

It is rare for a scientist to initiate a deep change in the understanding of nature, a so-called paradigm shift. Most scientific beliefs are quite stable because they are based on a matrix of interlinking information extracted by multiple experiments. For this reason, scientists are understandably reluctant to relinquish a stable understanding. Such a major shift in viewpoint is likely to be resisted by mainstream scientists, perhaps for years with the reminder that, "extraordinary change requires extraordinary proof." Scientists in search of the glory of establishing a paradigm shift should remember that the credit for a paradigm shift sometimes goes to those who later use more persuasive methods or have better publicized findings.

Many scientific experiments are designed to evaluate physical, chemical, biological, and social systems. The pace of discovery is so rapid that every day the results of scientific investigations across the world answer thousands of questions. Of course, hundreds of thousands of questions about nature still await answers. Scientists view unaddressed questions as splendid opportunities—opportunities that may elevate the quality of our lives through medical advances, improved crop yield and nutrition, the development of alternative sources of clean energy, more reliable and versatile communication, better housing, and so forth.

Blemishes on the face of science. Open and objective methods minimize but do not wholly eliminate imperfections in the practice of science. The social structure in science favors submissions from elite institutions and disfavors imaginative work from institutions chronically short of material support, as in Eastern Europe. Research also has its fads and fashions; at any given time, particular methods or areas of investigation are in vogue. Momentum and financial commitments may extend a research program beyond its usefulness, considering that new findings have already pointed the way to better research approaches.

An excess of personal ambition in science may lead to scheming to gain advantage, undercutting rivals, self-serving manipulation, abuse of power and privilege, lifting ideas or data from others, rushing to publish before the analysis is completed, bending conclusions to benefit the financial sponsors, failing to give credit to others while taking more credit than due, stymieing the efforts of competitors to publish their work, and ducking closer scrutiny of one's own work by publishing in a journal controlled by cronies. These are some of the human weaknesses that can blemish science.

Rare factual errors may elicit a correction published by the authors in the same journal. Sometimes, the entire article may be withdrawn as flawed. More frequently, the self-correcting features of science are less obvious—mistaken interpretations are not publicly acknowledged. Avoiding public admission, those who have been promoting erroneous conclusions or incorrect models will eventually subside into silence as the accumulating data persuasively sinks their views. Ultimately the truth will emerge. Whether credit for bona fide advances is fairly apportioned among different contributors is quite a different matter.

Unlike those able to use faith as a shield against uncertainty and ignorance, science must face up to both. Uncertainty is always present at the frontiers of scientific investigation where understanding is inevitably partial and provisional. Outright factual errors are uncommon but mistakes do occur. Scientific knowledge is like a freshly washed car. For the most part it gleams and sparkles, but there are always neglected areas that need further attention. These blotches are important if they reflect a crucial flaw in research or reasoning. Scientific knowledge is tentative; some revisions are to be expected over time.

Because science generally gets high marks for integrity and credibility, it attracts opportunists who exploit the lofty reputation of science. Park (2000) has distinguished four classes of "voodoo science." "Junk science" consists of investigations, which seem to offer definitive conclusions and tantalizing applications but turn out to be mere speculation lacking supporting data. "Pseudoscience" makes assertions with no scientific foundation—just a verbal parade of scientific jargon to lend an aura of credibility. Magnetic therapy is a pseudoscience that claims, among other things, that if you were to strap a small magnet to your arm, it would relieve the muscle aches and pains of a weekend of tennis (Flamm, 2006). Magnetic therapy is one of the alternative therapies that the U. S. Congress touted in 1992 when it established the Office of Alternative Medicines—now the National Center for Complementary and Alternative Medicine at the U. S. National Institutes of Health. The third class of voodoo science is "fraudulent science," where financial greed propels the commercialization of false claims. The fourth class is "pathological science" in which sincere but blindly enthusiastic scientists manage to fool themselves. The recent saga of "energy released by cold fusion" seems to be an example.

Some of the most pitiable complaints about science come from those hoping for immediate benefit as they overestimate the speed with which science is able to generate understanding and to frame workable solutions. Progress most often comes in small steps; science rarely makes giant leaps. Many scientific problems defy understanding, like predicting: earthquakes, tsunamis, volcanic eruptions, and the frequency of hurricanes; or curing schizophrenia, most degenerative neurological disorders, chronic kidney failure, and most forms of cancer. Even where there is understanding, it may be quite difficult to gain control. Scientists have long understood both atomic nuclear fusion and sickle cell anemia, but have yet to control either despite persistent efforts.

Commercial realities: Patients should be in the right place and in large numbers

Orphan drugs and profit calculations. The lure of profit spurs the marketing of drugs. Regrettably, it is hard to make much money from an illness that afflicts only a few people. The anticipated revenue would not cover the millions of dollars it takes to develop and get approval to sell a new drug. This means that candidate drugs for rare illnesses may languish like orphans hoping for adoption. In the pharmaceutical business the commercially ideal drug is expensive and protected by patents. And while it relieves a common chronic disorder, the "ideal" drug must be taken for life because it never really cures the ailment. To keep the profits rolling in, pharmaceutical firms have been known to downplay adverse reactions to a profitable drug and have been reluctant to develop cheap alternative drugs. Cheaper remedies would obviously lower profits, especially if they provided a lasting cure.

For several decades pharmaceutical firms took in billions of dollars selling antacid concoctions and patented inhibitors of stomach acid secretion that brought temporary relief to those with stomach ulcers. This generated a steady stream of profits because repeated flare-ups of symptoms required further rounds of expensive medication. As taught in medical schools worldwide, it was the universal view that the stresses of modern life caused acid secretion and gastric ulcers. However, in the 1980s Dr. Robin Warren, a pathologist at the Royal Perth Hospital in Australia, came up with the odd notion that gastric ulcers were caused by bacterial infections of the stomach. No one else took this seriously. For starters it was thought that bacteria couldn't even survive in stomach acid. Dr. Warren and a plucky young colleague, Dr. Barry Marshall, managed through intense effort and a measure of luck to grow stomach bacteria they named *Helicobacter pylori.*

In a dramatic and risky attempt to prove this bacterium could damage the stomach, Marshall swallowed a beaker of *H. pylori*. His resulting severe gastritis provided painful evidence for bacterial infection of the gut.

Today, cures of gastric ulcers are commonplace with the proper regimen of ordinary antibiotics. It is estimated that *H. pylori* causes 90% of ulcers in the duodenal intestine and 70% of stomach ulcers along with some forms of stomach cancer. This bacterial cause of stomach and intestinal ulcers is much more than an interesting medical discovery; it is one of the greatest upheavals in medical tradition and practice. With no support from other doctors, it required a prolonged and intense campaign of persuasion to overcome the inertia in medical training and the resistance of a pharmaceutical industry worried about the slim financial returns if cheap antibiotics really were able to cure ulcers. In 2005 Marshall and Warren were awarded the Nobel Prize in Physiology or Medicine for the scientific component of their efforts.

The geography of medical advances. Well-cared-for residents of Europe and the United States may believe that vaccinations have succeeded in preventing most of the serious microbial diseases, apart from those that are sexually transmitted. However, 40% of the world's population lives in areas where *Anopheles* mosquitoes transmit the protozoans responsible for malaria. *Plasmodium faciparum* is the most common malarial protozoan. Among the 400 million people infected, more than 2 million die every year—one death every 15 seconds (Poser and Bruyn, 1999). In the 1990s ten times as many children died of malaria as died in all wars during that decade. Unhappily, the *Anopheles* mosquitoes are becoming increasingly resistant to insecticides. Even worse, some mutant forms of *P. faciparum* are resistant to chloroquine and quinine, the traditional anti-malarial drugs. First known from the Ecuadorian cinchona tree, quinine has been the classic treatment for malaria. The growing hardiness of strains of mosquitoes and malarial protozoans highlight the need for an adequate vaccine. Now that the genome of *P. faciparum* has been decoded, there is hope a vaccine can be developed against stable portions of this malarial protozoan. At least malaria is no longer a neglected disease; some 47 malarial vaccines and 21 drugs are presently headed for evaluation at dozens of clinical testing sites in Africa.

Box 11.1 High Level Malarial Fevers.

Malaria takes its name from 'mal aria' or "bad air," which for more than 2,000 years was believed to cause malaria. If, as the Roman senator Cicero claimed, the gods were responsible for malaria, they unaccountably failed to spare the highest of the pious. Vatican City was built in Rome near mosquito-infested wetlands. The fatalities were appalling. In the 13th century alone 17 popes died of malarial fevers. During the Vatican conclave of 1623, malaria killed eight cardinals and thirty secretaries. Infected Spanish conquistadors had already spread malaria to the New World in their journeys across the Atlantic Ocean in the 1500s. Soon thereafter infected Africans bound for slavery were additional carriers to the Americas. In Africa malaria had been a human scourge long enough for a genetic defense to have arisen naturally. A particular variant of the hemoglobin gene causes human red blood cells to adopt a sickle shape. This results from the substitution of just one amino acid in the hemoglobin protein. Because both mother and father contribute chromosomes, every cell has two copies of most genes. The combination of one copy of the sickle cell gene and one normal copy partially distorts the red blood cells while reducing the rate they are infected by malarial protozoans. But if both hemoglobin genes are of the sickle cell type, the highly deformed red blood cells may so seriously interfere with blood flow that the "sickle cell anemia" is fatal.

Commercial Greed and Politics Can Damage the Process and Application of Science

The natural world operates by mechanisms and rules that scientists are often able to discover. Scientists neither construct the rules of nature nor establish them by decree. Scientific advances turn on data, on new empirical findings that hold up under continued scrutiny. In contrast manipulation of data and conclusions is prevalent in the worlds of profit and politics.

The ruse of homeopathic medicine (Park, 2000). In the early 1800s Samuel Hahnemann, a German physician, espoused two medical principles: "like cures like" and "the law of infinitesimals." As a young man he recovered from malaria with a dose of quinine that gave him malaria-like fever and chills. This solitary experience was enough to convince him that a natural substance would be an effective remedy if it produced symptoms like those of the disease. This so-called homeopathic medical cure still relies heavily on a list of natural substances that Hahnemann adopted based upon side effects that mimicked the disease symptoms. To reduce the severity of a potion's side effects, Hahnemann frequently diluted his chemicals to infinitesimal concentrations; like one molecule in 8,000 gallons! In order to put a "scientific" spin on homeopathic therapy it was admitted that even though the dilution left few if any drug molecules remaining in solution, the water must have retained a "therapeutic memory" of their former presence, whatever that speculation might mean. A leading proponent of homeopathy, Wayne Jonas, Director of the Center of Alternative Medicine at the National Institutes of Health, has coauthored the book *Healing with Homeopathy* (Jonas and Jacobs, 1996). Since an itch deserves an itch, these authors recommend treating diaper rash with poison ivy. Not to worry, the recommended poison ivy extract was almost as dilute as water. Because of an ill-advised congressional exemption from Federal Drug Act regulations, ever since 1938 homeopathic remedies have dodged review for effectiveness and safety. After all, it is only a watery placebo, so safety is not an issue, as long as one ignores the consequences of foregoing truly effective remedies to treat an illness. Homeopathic medicines may be one of the most lucrative forms of the placebo effect yet devised; it is pseudoscience in grand form, effectively selling water as a cure-all.

Politically motivated distortions of basic scientific research. As one recent example, in the early part of the 21st century the United States executive branch has repeatedly undercut and censored scientific evidence that conflicted with its political agenda. For example, when presented with recent strong evidence for global warming, a powerful government lawyer systematically censored and altered the findings and conclusions of these scientific reports to downplay the warming trend. Multiple press releases from NOAA (National Oceanic and Atmospheric Administration) were actually blocked because they documented global warming. Unfortunately, global warming does not stop merely because a government buries its head in the sand. The Arctic ice cap is melting faster than previously anticipated. There remains an ocean of uncertainty over whether governmental dithering will make it too late to prevent catastrophic global warming and related climate changes which are unequivocally driven by human activities.

Political faking of technical results. Open access to information and the transparency of behavior are among the best hopes for keeping people honest. Fortunately, in most circumstances, the scientific process is open for review. However, when corporate or military secrecy thwarts scrutiny by neutral observers, it can corrupt assessments. Although initially highly praised, the ability of the Patriot missile to intercept Iraqi Scud missiles was repeatedly downgraded after the close of the first Bush's war with Iraq. Eventually it was admitted that there had been no confirmed Patriot interceptions of Scud missiles at all. More recently, the United States Navy declared its new air-to-air missile had a successful launch, even as they failed to disclose the missile soon went haywire because it got so hot it fried the on-board electronic circuitry.

One of the most dangerous things craven politicians and the heads of the military-industrial complex can do is claim that they have an impenetrable missile shield. If politicians and naïve citizens become confident that we are invulnerable to nuclear attack, the public and their leaders might be sufficiently emboldened to fight Pakistan or China or Russia over possessing nuclear arms. The potentially disastrous consequences of a failure of our so-called shield are beyond ordinary comprehension and surely beyond medical help. Should even one, just one, hydrogen bomb explode above any medium sized city in America, it would probably vaporize the skin of hundreds of thousands of children (Thomas, 1983, p. 118). There is a simple and universal rule in all fields, including commerce, government, science, and religion—conclusions generated by secret decisions are not credible because secrecy so frequently corrupts objectivity.

Science and Religion Compared

Mysteries. Science and religion differ strikingly in why they are attracted to mysteries. Science seeks immediate mysteries to explain; religion seeks perpetual mysteries to sustain. Religious faith targets the distant future. Science emphasizes present conditions and how they arose. Of course, in an era of global warming and environmental deterioration, science is also actively grappling with the long-term future. Still, biological science is more able to understand the past and explain the present than to predict the distant future. The uncertainties of the future provide strategic opportunities for religion to offer optimistic promises, notably future supernatural rescue operations beyond the reach of science. Uppermost is the final fate of dying humans. Although religions stake out a claim on the future, specific predictions have frequently failed. One loses count of the number of times Jesus was due to return, or the specific days the world was slated to end. Today, such claims risk comic ridicule and tragic outcomes.

Science and religion also take different approaches to the search for answers. The devout may already know in advance what answers and outcomes they want. Christians generally want an all-powerful God of mystery who responds to prayer on Earth and, through Jesus, offers Salvation in Heaven. Pre-committed to such outcomes, they then seek reassurance from the available evidence, from pastoral and scriptural teachings, and ultimately from feelings and faith. Scientists elect to do experiments on some particular structure or process to discover its features and functions. Rather than needing a particular outcome, scientists strive for clear-cut results that lead to unambiguous conclusions.

Power. One can expect theologians to frame controlling doctrines—allowing a menu of choices or optional beliefs is scarcely popular among preachers. What is an industrious shaman, or a prophet in the Holy Land, or a preacher in the heartland striving for if not to exert control over a collection of people? Much before the phenomenon of Protestant mega-churches, success in religion meant corralling expectant crowds. An ideal objective would be to lift people's spirits and outlook with homilies of hope and fellowship. Theological liturgy is laced with "stick and carrot" motivators—tools that empower doctrines. Attempts to gain compliance by fueling insecurities with fire and brimstone messages about Hell is abusive control—unless you're convinced the recipient truly needs such threats of eternal damnation.

Trinitarian Christianity asserts its power through its stupendous God without limitation, able to enter the hearts of the faithful as the Holy Spirit, and transformed two millennia ago into a suffering, human-like Jesus. As *John* 14:9 put it succinctly, one who has seen Jesus "has seen the Father." The commitment to maximization is one of the most striking aspects of Christian theology. God is all-everything, stupendously more than even all-universe which, after all, he created. How do we understand the repeated theme that God has limitless powers and capacities? Christians want God to be the greatest of mysteries, colossal in size, age, power, and love. On the other hand Christians also find that the manly image of Jesus helps them identify with God the Father. To claim God is both the ever-present Jesus and also a much more expansive Father figure leaves outsiders quite perplexed, if not slack-jawed with incredulity. Moreover, Jesus never claimed to be divine—humans raised him up to it.

A life of surprises. Science provides the understanding that may then lead to technological inventions and hoped-for cures. The theologian John Haught (2007) claims that it is religion which furnishes surprises, whereas science merely explains the already expected. But the reverse is true. Since religion generates little new knowledge, it is no surprise that most theological arguments have grown stale after centuries of indecisive debates. In contrast, surprise is a core trait of science whose journals are filled with novelty. There are many keen surprises in chemical interactions that can lead to startling outcomes. In a classic example, from a combination of two parts hydrogen gas and one part oxygen gas, no one would have been able to predict the many remarkable properties of water as the product. Biological research is also generating major surprises at a remarkable pace. It came as a total surprise that there are large families of short RNAs, which influence gene expression in multiple ways. And who would have thought that lowly sea urchins would have genes previously believed to be strictly mammalian? Or that we use hundreds of genes just to detect foods by their smell, or that nerve cells in the brain communicate with one another by releasing small amounts of the deadly respiratory poison carbon monoxide. It would be quite fitting to entitle the biography of most biologists, "A lifetime of astonishing discoveries."

Future worship. We like our earth, our galaxy, and our universe. We don't want to admit they are all ultimately doomed—destined eventually to disintegrate and disappear. Make no mistake, immortality matters to humans. Those seeking true immortality will refuse to settle for a mere several billion years of afterlife while our universe plays itself out. Their afterlife must go on and on—indefinitely. Since this could require triumph over the finite life of our universe, some theologians look for renewal, or more reassuringly an infinity of renewals, universe after universe.

Theologians commonly disparage the meaning and significance of our brief secular lives on Earth, thereby enhancing the apparent significance of eternal Salvation. As the Catholic theologian John Haught put it, "Theologically speaking, nothing finite could be purposeful, unless it partakes of the eternal." (Haught, 2007, p. 54). How is it that there is purpose only in the eternal, the infinite? Is immortality so paramount that we demand a perpetual succession of replacement universes? Why should an ending so staggeringly remote in time be taken as a sign that your life today has no purpose? And where is an accessible Heaven that its occupants will survive long after our present universe has lost its form and energy? Are they miraculously whisked away to a Heaven in another universe billions of light-years away? Surely some degree of realism must seep into these fantasies. To nurture our children and family, to form supportive communities, to provide for the needy—none of these has a purpose unless somehow linked to the eternal? Isn't this numbing theological speculation just an all-out obsession with defeating death? It is a pitiable, yet theologically encouraged attitude, that immortality is our only true purpose—that the multitude of brief lives on Earth are meaningless.

Let's consider a single-celled organism that must have lived long ago. We could identify it by its number in the lineage of cells (like #1,532,870). Assume it reproduced just before it died and disintegrated—forever dead and gone. Now before you say this brief cellular life was meaningless you should consider one possibility. This may be one of a long series of cells—an early link in the unbroken chain of life that millions and millions of years later led exactly to you. Now do you think that the health and welfare of this cell was inconsequential? Biological evolution is the narrative of billions and billions of such brief fleeting lives, diversifying radiations, most of them dead ends, but others undeniably essential to their human descendants, such as you.

A Delicate Détente with Religion

Today's religions have the demanding task of sustaining their traditional core beliefs while at the same time avoiding conflicts with science. One popular way to clear room for God and supernatural influences is to frame doctrines beyond the reach of science. The theological elements most likely to evade scientific scrutiny include the existence of hidden or camouflaged actions of God to introduce immaterial human properties like our souls, original sin, personal communication with the divine, and the value of specific rituals like baptism. For many denominations the watery immersion of baptism aims to invisibly infuse each person with the Holy Spirit and wash away original sin. Even if baptism led to an observable change, it would require access to the supernatural world to prove a supernatural cause. No one has yet offered credible methods to access the supernatural realm. Because science is limited to the material realm that can be observed or measured, invoking a supernatural influence is not a scientific explanation. Notwithstanding Dante's superb imagination, scientists will never measure the daily temperature of Hell, or determine whether Hell's torments are worse than present day secret tortures carried out by 21st century authoritarian and democratic nations. The only access to the properties of Hell is through the minds of humans motivated to imagine such a hideous place littered with terrifying features. The threat of Hell is intended to intimidate sinners and potential apostates.

The drastic fall-off in credible claims of miracles and prophecies in modern times is the result of closer scrutiny. As if we needed more proof of the resilience of spiritual strivings, increased scientific scrutiny of religious myth has often heightened, rather than diminished, the yearning for signs of divine intervention. Emergent signs of God are found predominantly in visual oddities like a coffee stain or a cliff's mossy covering whose shape resembles the face of Jesus, or in claims of spiritual encounters, dreams, near-death experiences or luminous enlightenment inaccessible to others. Church authorities discourage the populace from announcing a miraculous sign, like “tears” on the face of a statue of Mary. Clerics don't want to be the bearers of deflating news that the most recent “miraculous sign” they were pressed to investigate is yet another matter-of-fact example of ordinary chemistry or human cunning. The far-reaching ability of the scientific enterprise to understand and control the physical and material world allows it to solve puzzles and make many mysteries disappear. It's no wonder that present-day miracles are difficult to come by.

The embarrassment in making a public retraction ensures that public confessions of religious errors are both rare and late. Institutional admissions of wrongdoing may take centuries. Still, it is a sign of integrity if institutional leadership eventually comes around to admitting its moral shortcomings. For example, during his papacy, Pope John Paul II made more than 100 apologies for various historical wrongs of the Roman Catholic Church. These included apologies for passivity during the Holocaust (March 16, 1998) and for the unjust treatment of women (July 10, 1995)—an issue that still festers. It was a bleak October day in 1992 when the pontifical commission and Pope John Paul II lifted the Vatican's condemnation of Galileo's astronomical observations—an apology that must have been gratefully received by Galileo's descendants three and half centuries later. Nowadays one doesn't expect religious documents to provide specific working details of Heaven or of God for this would risk putting their features within the reach of science. (But see Chapter 15 on “Intelligent Design.”)

Will science ever carry the light of understanding into the murky realms of supernatural beings, the soul, immortality, or Heaven's features? Not that the natural realm lacks its own puzzles that defy solution. Within neuroscience, the nature and causal capabilities of the conscious mind are the paramount puzzle that defies scientific access. If it could be shown whether and how mere immaterial thoughts can influence brain physiology, it might hint

at how the supernatural could influence the material wo
both situations imply an action of the immaterial on the
As Lord Adrian noted years ago, biologists have never mana
to capture "material thoughts" the way one might collect drops of saliva in a bottle.

It is difficult to understand the claim of mind-body dualists that networks of nerve cells in our brains give off immaterial thoughts—like a gas without molecules. It is even more difficult, perhaps impossibly difficult, to understand how such an immaterial thought or idea could arouse material nerve cells to produce speech or arm movements. In rejecting such dualism most neuroscientists believe that mental experiences (thoughts, feelings, sensations, etc.) are aspects of the physiological activity of networks of nerve cells. This must be the case if mind and brain are one. However, this hardly clarifies the matter—in no sense do we literally "hold that thought." Since it is so counter-intuitive that a sensation like the color yellow or anxious thoughts about the location of a misplaced credit card are nothing more than electrical activity in a set of nerve cells, this has been called "the astonishing hypothesis" by Francis Crick, Nobel prize winner for the co-discovery of the molecular structure of DNA. The issue here is not complex information processing but rather the direct simplicity of some mental experiences—real but immaterial sensations. At the beginning of the 20th century psychologists busied themselves trying to experience the most simple and elementary sensations, unadorned with knowledge and past experience. Banana odor, a clear blue sky, a pure tone, and light touch are such mental experiences. We know these sensations come and go with specific changes in the state of the brain, but does that explain anything for you?

Do Religion and Science Conflict?

The answers to this mega-question depend on the reach and scope of each of these two domains. There is no demonstrable conflict between science and deism, which asserts that God started the universe and has not intervened since then. But science refutes the claim of Christian Fundamentalism that the Creation accounts found in biblical Genesis are literally accurate. More generally, modern scientific findings conflict with many earlier religious assertions about the character of the material world.

A strained relationship. In one common religious metaphor the tensions between science and religion resemble a strained marriage. Certainly there is a long history of stressful interactions between scientific and religious authorities. Like a wedding between a Greek mortal and a god, it is an unnatural pairing—science deals with the natural world and religion with the supernatural. But religions have always wanted the supernatural to intervene in human life through relief by prayer, divine guidance, received blessings, baptismal transformations, inspired prophets, and more generally the holy direction of life on Earth. Innumerable Christian prophets and theologians have speculated about the actions of multiple supernatural figures. Roving bands of angels were said to live in a firmament of celestial spheres. Satan and his execrable companions lived deep in the bowels of the earth. While on the earth's surface, God punished disbelieving pagans with plagues and pestilence. These frequent supernatural intrusions by righteous idols and merciless demons are worn out images. Nonetheless, even today, arrogant preachers embarrass thoughtful Christians by calling down God's wrath on skeptics, as if God will carry out a preacher's malicious directions on command.

Science does bad things. It is a bogus complaint that the rise of scientific materialism (the belief that there is no supernatural realm) has devalued life. This public relations gambit is part of a counter-reformation against science intended to diminish its stature and elevate institutional religion as the only true standard-bearer of life. Biologists of my acquaintance need to yield to none in their reverence for all manifestations of life (Goodenough, 1998). From the brilliant interference-colors of a butterfly wing, to the intricate linkages among hundreds of biochemical pathways, or the harmonics of a nightingale's song—biologists revere life. Others may cling to the idea that living cells and organisms are infused with a vital spirit. Yet, as the parts of cells become increasingly understood, biologists see no need to add a splash of "vital spirit." The component parts of cells may soon be combined artificially to construct a living cell (see Lartigue et al., 2007). What are the differences between a dead cell and a living cell? The main differences are that a living cell can digest food molecules and use the released energy to synthesize other molecules for growth and maintenance. Living cells may reproduce.

Beyond scientific understanding. Some may resent the advances of science that shrink the scope of religion whose cherished myths once fed our spiritual needs. As scientific understanding grows, the confrontation with religion is being played out on a shrinking chessboard having fewer and fewer squares of mystery where theologians place their pieces. If scientific understanding were to purify religious beliefs and retire erroneous myths, would it fatally enfeeble religion? My sense is that although the space for faith is shrinking, it will never disappear. The life-blood that sustains religion is the set of impenetrable mysteries and insoluble questions that focus on our ultimate purpose and fate. Scientists work hard to frame a problem in a sufficiently precise and narrow way that proposed answers can be tested and rejected according to the data. Religion on the other hand embraces ultimate questions that won't yield up answers to science.

Separate Domains for Science and Religion?

Before the advent of science, religious beliefs characteristically helped to fill voids in human understanding. When science arose as an independent approach for developing credible understanding, it began to compete with religion for explanations of natural phenomena. As the facts of science have accumulated, the frequency of fanciful "explanations" has subsided. By incorporating scientific findings, contemporary religions avoid conflicts with science. It may be prudent to "make no claim within the reach of science."

In an essay entitled "Dorothy, It's Really Oz," Stephen J. Gould (1999) wrote, "Science and religion should be equal, mutually respecting partners, each the master of its own dominion, and with each dominion vital to human life in a different way." That is, science and religion are parallel and similarly magisterial enterprises overseeing separate domains. They must respect each other and co-operate in examining reality. Gould offered such observations with the hope of minimizing future skirmishes. Unfortunately, respect arises from mutual agreement, not from a third-party request to harmonize and make peace. In denying conflicts between science and religion, Gould was effectively recommending the prudent strategy that religion should stick to existential issues that are inaccessible to science. However, historically religion has exercised a broad dominion over the full range of life, at least until the expansion of scientific discoveries caused religions to scale back. Most scientists would feel uncomfortable approaching religious figures with suggestions for changing venerable doctrines to incorporate recent scientific findings. In contrast, confronted by nettlesome scientific advances, some clerics and devout laypersons spend significant time and effort in disparaging sticky scientific findings like Darwin's theory of the biological evolution of life that can act as flypaper for spiritual discontent. Indeed, how reasonable is it to expect the clergy or any thoughtful person to avoid commentary about the origins of human behavior and the nature of human character? Perhaps the only theological silence will be among those monks and nuns withdrawn into lives of contemplating the supernatural. In spite of Gould's hopes and some clergy's preferences, science and religion will continue to collide. As we will see there are ongoing clashes between religion and science over several features of human reproduction and development, most heatedly over the descent of humans from apes. For better or worse, science and religion do not have a reciprocal relationship. There is no rational basis for dismissing scientific facts on the grounds that they conflict with religious doctrines. This isn't to say that the scope of science and its methods haven't been reined-in by government regulations responsive to the concerns of organized religion. This can be a contentious matter in a civil society where the separation of church and state intends that specific religions do not impose their particular dogmas. Religious creeds and doctrines generally adhere to stable stances born from pre-scientific traditions. Indeed, part of the appeal of ancient scripture is to provide trusted eternal verities. Quite aware that scientific advances are continually clarifying reality, the stewards of contemporary religions generally avoid assertions that risk future retraction. As shown in Chapter 15 on Intelligent Design, it is risky to find God in the unknown—to create a "God of Gaps." When science subsequently fills gaps in human understanding, it can marginalize a sacred doctrine. When scriptures serve as sacred poetic metaphors rather than literal declarations, it can reduce conflicts with science.

On the Matter of Human Life

Abortion. It may surprise you to know that the Bible, including its more than 600 laws of Moses, does not set down rules about abortion; indeed the word "abortion" does not even appear in the Bible. Are we to take the absence of the word "abortion" in the scriptures as a sign of theological indifference? Hardly. From the earliest Christian times numerous theologians offered strenuous objections to abortion induced at any developmental stage. Many others considered abortion a sin only if it was performed after "quickening" when a mother begins to feel movements of the fetus. The most decisive consideration was the timing of "ensoulment." Those who expected the soul to enter at the moment of conception were among those opposed to abortion at any age. Those men rejecting all abortions included Clement of Alexandria (?-215 CE), Tertullian (155-225 CE), St Hippoplytus (170-236 CE), St Ambrose (339-397 CE), St Jerome (342-420 CE), and Pope Steven V in 887 CE. St John Chrysostom (ca. 340-407 CE) must have been in the throes of intense feelings when he remarked, "You see how drunkenness leads to whoredom, whoredom to adultery, adultery to murder; or rather something even worse than murder. For I have no real name to give it, since it does not destroy the thing born but prevents its being born" (*Homily 24 on Romans*). Also consider how St Jerome cast his saintly net to catch multiple sins. "They drink potions to ensure sterility and are guilty of murdering a human being not yet conceived. Some, when they learn that they are with child through sin, practice abortion by the use of drugs. Frequently they die themselves and are brought before the rulers of the lower world guilty of three crimes: suicide, adultery against Christ, and murder of an unborn child" (*Letter 22:13*).

Less restrictive was Aristotle (384-322 BCE, *Book VII*) who promoted the idea of "delayed ensoulment" (40 days after conception for a male and 90 days for a female fetus). St Augustine (354-430 CE, *On Exodus,* 21, 80) sided with Aristotle while disagreeing with many of his stricter clerical contemporaries. Similarly, St Thomas Aquinas (1225-1274 CE), in following the lead of Pope Innocent III (?-1216), believed abortion was not murder unless the fetus was "animated." These were the considered views of three A-rated scholars: Aristotle, Augustine, and Aquinas.

In his papal Bull *Effraenatus* of 1588 Pope Sixtus V threatened excommunication and death for those committing an abortion. Nonetheless, only three year later Pope Gregory XIV revoked this papal Bull and reinstated a quickening standard for abortions (116 days beyond conception). Presumably the medieval faithful were true to their training and dutifully lined up behind the last authority to speak. More recent revisions of Catholic Church Canon law in 1917 and again in 1983 have settled the matter for those who follow papal authority. Excommunication is the penalty even for early abortions. What has biology to offer on the matter of abortion and "human life"?

Pre-emptive losses. Those who anguish over abortion need to consider the sporadic reproductive behavior of women and men. Every man's missed opportunity to inseminate, and every ovulating woman's failure to have timely unprotected intercourse, potentially deprives the world of a new human being. To neglect any opportunities for procreation thwarts the natural union of egg and sperm. As an undeniable consequence of this neglect, such a preconceived child's one chance at life slips away, forever gone. This broad concern for human reproduction goes beyond merely obeying a religious command to have unprotected sex. It is a matter of faithfully attempting to prolong the life of each egg as it presents itself. Shouldn't a woman, who might possibly sense an egg descending down the fallopian tube, be quick to mate with her man for his sperm? This is the only natural way to prevent the death of this living egg, this pre-Susie. The reality is that most people decline to "choose life" for virtually all of their eggs and sperm. Every month an egg comes calling, but few are chosen. For every ovulatory cycle in which couples knowingly fail to make multiple efforts at procreation, they are engaging in pre-emptive abortion, allowing the egg to die. Those who have been repelled by contraceptives that prevent ovulation or entrap swimming sperm should consider whether they are living up to their principled stand in light of these fuller realities of the human reproductive system.

Human life requires a functioning brain. It is universally accepted that conscious human life requires substantial brain activity. For this reason, the medical profession defines the end of life as the absence of electrical activity in the core of the brain. This issue is highlighted by a horribly failed attempt to free a worker trapped in a caved-in ditch in Michigan. While desperately digging, the backhoe operator accidentally decapitated the construction worker who mercifully had probably already suffocated to death. In this most unpleasant ending to a construction accident, only a wholly distraught person would have wanted to prolong the metabolism of the worker's body on the basis that his headless corpse was not dead. It is true that even after the loss of brain function many tissues continue to function for a while, even long enough to be transplanted to a needy recipient. But it is the status of the brain that controls our assessment. A body without a brain is no longer a person. You can't have conscious life if you lack a brain. Imagine if it were possible to transplant your brain into another person's body. It is your personality that would survive, perhaps now housed in a young, well-toned body. If we had an injured spinal cord, we might be willing to accept donor tissue to replace the damaged section if it ever became technically feasible. But no thoughtful person believes his own life would continue if his body received a different brain transplanted from someone else. By the way, don't hold your breath; a mammalian brain transplant is a daunting technical feat and is not even a goal.

The development of the brain and the beginning of a person. It is helpful to review briefly some facts about human embryonic development. Following on from the single swimming sperm that successfully penetrated a viable egg at fertilization, every cell of the growing human embryo is alive. As young cells grow bigger, many of them divide. (It is a perfectly normal aspect of embryonic development that many surplus cells die, including about half of the brain's cells, as well as cells in the foot that die and drift away in helping to sculpt toes from a simple paddle. But the detailed contributions of cell death to development is a matter for textbooks.) Interestingly, until it is two weeks old the human embryo has no nerve cells of any kind. Not one. After nerve cells arise they must form connections with one another and still later become electrically active in a way that allows functions like processing and storing sensory information and issuing "commands" (Spitzer, 2006). Neuroscientists do not know how many nerve cells and what specific brain circuits are required for particular functional capabilities. They do know that the developing central nervous system (brain and spinal cord) of human embryos does not transmit its electrical messages before six weeks of age. The brain continues to enlarge until by 28 weeks a fetus has a brain capable of mediating breathing. Therefore, we can focus upon the functional changes during the important 6-28 week period. By 13 weeks nerve connections cause jerky skeletal muscle contraction and movements, but it is not until 23 weeks that a leg automatically withdraws from a pinprick. Since it takes relatively few connections in the spinal cord to mediate withdrawal reflexes, neuroscientists believe that a withdrawal reflex is too elementary a function to serve as a proxy for conscious awareness. Nonetheless, it is appropriate to be cautious and conservative about the development of neural circuits and to consider even elementary reflexes as potentially important. So it is reasonable to hold that at several months of age a fetus may be an individual, a person—someone with a brain that is at least starting to gain significant function, even if they are not awake and conscious.

To summarize, at two weeks of age the embryo is an inconspicuous hollow ball of non-neural cells smaller than the period at the end of this sentence. If the loss of brain function marks the death of the individual at the end of life, then the life of an individual person cannot begin until there is brain activity where nerve cells have become linked together in functional circuits. The egg's encoded plan for construction of a brain is not the same as actually having a brain. The most that can be claimed is that the potential for one or more persons is created when the sperm fertilizes the egg. (See the discussion of twins below.)

Implications for the soul. Scientific findings can spark theological reassessment of the natural world and the supernatural realm. Prior to the invention of telescopes, Christian theology saw the earth's moon as unique. Modern telescopes and satellite reconnaissance have since revealed additional moons revolving around several other planets. Such information requires a rather inconsequential change in the theological view of the natural world. In contrast, any abrupt adjustment in supernatural beliefs will be problematic. Revision of a supernatural trait prompts one to ask, "why was it that the earlier supernatural trait was incorrect, and how do we now know the latest choice is the correct one among the alternatives?" A few examples follow.

In the development of a human, it is not uncommon that an egg is fertilized, but fails to develop further. Indeed, in nearly half of human pregnancies the embryo is promptly reabsorbed or sloughed off. It is difficult to see how this frequent failure to develop, this spontaneous abortion, can represent a person's death or the loss of a soul when there are no nerve cells. It is a lost prospect, rather than a lost person. If a single cell, a fertilized egg with genetic plans for construction is already a human being, then by parallel reasoning a box of blueprints is a Hilton Hotel, even before construction begins. One retort is that this is a flawed analogy because no one claims that blueprints have a soul.

Even though the soul is defined as immaterial and invisible, weightless and lacking an electrical charge, it is not beyond the reach of science to set some limits on its claimed actions. Traditional theology says every person's soul is unique. Even so, if one of your skin cells were cloned, what is its poor soul to do? Theologians forbid you and your new clone from having the same soul, but there it is. Or does the skin cell's soul die and give way to a replacement soul unique for the clone? Who decides how to recast such problematic features of the supernatural? And how do we know this is how God actually recasts the supernatural? The problem is also present in identical twins or natural clones.

It is possible to test the claims that one soul enters at the moment of conception (when the sperm penetrates the egg) and is united with the egg as one embryo until death. For the Catholic Church ensoulment at fertilization has been a long-disputed article of faith, since there is no way to detect if and when the soul enters. How does ensoulment work out with identical twins? You may have friends or acquaintances who are identical twins. Identical twins have identical genes because several days after fertilization an embryo duplicates its DNA and then splits into two identical embryos that develop independently. Since this split characteristically occurs more than one week after fertilization, we need to ask whether the second soul also entered at conception. This would mean that from fertilization until the moment when the embryo splits there were two souls residing in one embryo. But this would be contrary to a fundamental tenet—that each person, each body, has but one soul. Nor would it be satisfactory to have one soul split into two souls when the embryo splits, for each human soul is unique—identical twins have identical genes but different souls. So is it with some delay after conception that the second soul or both souls materialize (if that is the word) as the embryo splits into two, so that there is only one soul per embryo? This suggestion is complicated by an additional twist to the problem of twinning. Even after there are two embryos or twins, each with its own unique soul, it sometime happens that these separate embryos fuse back together again into one embryo. What happens to the second soul, when again there is but one embryo? Is the second unneeded soul simply cast out? What happens to a soul that enters at conception only for the embryo to be spontaneously aborted after a few days? Are all doomed embryos kept soulless because God foresees, but does not prevent, the loss of the embryo? There are many detailed questions about souls, but few ways to approach them scientifically. Nonetheless, as I have described, biological studies of the timing of splitting into identical twins, occasionally followed by re-fusion, have revealed that at least in the instance of twinning, the biological facts require exceptions to the claim that only at fertilization is its one and only one soul introduced into the egg along with the sperm. Doubtless church authorities could respond with exceptional scenarios, but who today has such a privileged source of information about ensoulment as to know with confidence not only of the soul's existence but also of God's general rules for the behavior of souls and even the specific exceptions now required by better biological understanding? By far the simplest general account of soul behavior consistent with biology is that a soul does not enter the embryo until several weeks post-conception as per St Thomas Aquinas. Around this time the advent of a working nervous system provides the basis for a specific individual with no possibility the embryo will either split into identical twins or later fuse back into one.

In such evaluations it is important to distinguish between plans and projects. Imagine that a lightning bolt hits an assembly plant just as the aluminum roll that was destined to be your uniquely personalized $100,000 mobile home was starting down the totally automated assembly line. Should the insurance pay to replace just the roll of aluminum, or an entire finished mobile home that was soon going to be built automatically according to plan? Analogously, as we think about a morning-after pill we must distinguish between the starting material with its plan to be read out, and the finished product constructed over many weeks precisely according to plan.

Additional Collisions between Science and Religion: Past, Present, and Future

Faith-based dogma has a history of being flat out wrong—the earth is ancient and far from the center of the universe. Our most recent ancestors were apes. Some of our vital genes are remnants of the DNA left by microbial parasites. Are we to deny these things and specify what is correct simply through our faith in the most recent declaration of faith?

A brief look at Christianity and the theory of evolution. The Catholic Church and numerous scientists who are Christian believe that Christianity and evolution are compatible. Even so, the implications of evolution loom above all other scientific findings as the most tormenting to many conservative Christians. The distress begins with a core factual disagreement among accounts of Creation. Biologists believe that all organisms are genetically related through the continuity of descent. This view conflicts with literal interpretations of Old Testament scripture and the equivalent portions of the Hebrew Bible and the Qur'an. These are difficult times if your life is tethered to the acceptance of scriptural literalism. A variety of radioactive dating methods prove that the geological earth and the life that soon followed are billions of years old. Moreover, there is conclusive scientific evidence that multicellular life evolved long ago from single-cell organisms. If living organisms are closely related they should share numerous genes. Indeed, as shown by recent studies in molecular genetics, the relative proportion of shared genes is a useful predictor of relatedness.

On October 22, 1996 Pope John Paul II told the Pontifical Academy of Sciences that, "Masses of evidence render the application of the concept of evolution to man and other primates beyond serious dispute." This was a re-affirmation of the papal analysis in 1985 that accepted the fact of the evolutionary descent of organisms but disputed the mechanism. The papacy concluded that species were divinely created—that each successive species tumbled from the hand of God. (Evolutionary biologists have rather different views about speciation.) Pope John Paul II further asserted that although the human body evolved from apes, God installed the human spirit (mind). He said that the human mind is property that naturally emerges from greater biological com ity nor a mere epiphenomenon of neural activity. Although many have noted the thoroughness of Pope Paul II's assessment in concluding that evolution is compatible with the Catholic faith, the newly appointed Pope Benedict XVI may have a different reckoning. In 2004 the Catholic International Theological Commission headed by the now Pope Benedict XVI seemed to conclude the Catholic Church had no problem with neo-Darwinian evolution. Yet, it also stated that, "An unguided evolutionary process—one that falls outside the bounds of divine providence—simply cannot exist." More recently in July of 2005, Cardinal Christoph Schönborn disparaged Pope John Paul II's October 22, 1996 supportive statement on evolution "as rather vague and unimportant." Cardinal Schönborn championed the view that all reasonable people see designs in nature that offer sound evidence for God. He remarked, "Any system of thought that denies or seeks to explain away the overwhelming evidence for design in biology is ideology, not science." Biologists respond that the mere existence of complexity is not proof of divine design. Evolutionary principles can explain biological complexity. The Vatican has since backed away from Cardinal Schönborn's statements. Chapter 15 will evaluate "Intelligent Design" as evidence for divine intervention in life.

Scientists can be forgiven if they envy the breathtaking ease with which the invocation of supernatural intervention explains life's complexities. This option is out of the reach of science, which is limited to hard-won measurements in the natural world. Scientific explanations for evolution must be framed in mechanisms that deal with material events—evolution by processes and principles that can lead to testable predictions with measurable outcomes. By invoking the divine, theologians circumvent having to commit to scientifically testable assertions. No less daunting than the ancient challenges set before Hercules, religion's perennial challenge is to sustain religious beliefs without running afoul of scientific assessments of knowledge. This is particularly evident when the properties of venerated religious icons are assessed with scientific techniques.

The winding cloth of the crucified Christ. Science is sometimes able to evaluate claims of supernatural interventions. These claims may take such forms as weeping statuary, divinely inspired prophesy, clairvoyance, mental telepathy, or claims that a venerated relic is authentic, like an image of Jesus transferred onto a burial shroud. Such claims can be tested for a natural explanation that doesn't require supernatural intervention.

In modern times radioactive dating methods have been used to check the actual age of Christian religious relics. Consider the history of the Shroud of Turin, claimed to have received images transferred from the front and back of Christ while his body was briefly wrapped up after his crucifixion. Science and religion have had contentious exchanges about the authenticity of this venerated winding cloth (Nickell, 1998).

In Europe during the Middle Ages at least 44 different cloths vied for the honor of having wrapped the body of Jesus. One of these cloths (later called the Shroud of Turin) first appeared in a small church in Lirey in north central France, where it was put on display in 1355 by the soldier of fortune Geoffrey de Charney. By 1389, the local bishop had concluded the cloth was a painted fraud being used to scam money with the help of two "crippled" confederates who would scamper about feigning recovery at the instant the shroud was unfurled before a gullible crowd. Pope Clement VII was soon persuaded to declare this shroud could be displayed only with the disclaimer that it was a painted representation of the real shroud. Still, the papacy must have attached some religious or political importance to the shroud since De Charney's granddaughter was excommunicated in 1457 after her refusal to give up the shroud to the church. For the next five centuries the Dukedom of Savoy and the Italian monarchy kept the shroud until it was eventually bequeathed to the Vatican in 1983. Although the Catholic Encyclopedia of 1912 concluded the shroud was not authentic, religious officials continue to fan the hope in the hearts of parishioners.

In 1969 and again in 1973 the Archbishop of Turin appointed secret commissions to evaluate the authenticity of the Shroud of Turin. As the most venerated relic in Christendom, any public attempt to prove it was authentic was certain to arouse intense interest. Access and sampling were strictly controlled. In spite of extensive efforts by those skilled in the chemistry of blood, their tests revealed only traces of red paint. No blood could be detected. The finding of red paint was announced and then quickly suppressed while the church proceeded to freely circulate a rebuttal of these uninspiring findings. This was to be a repeated pattern—after distinguished scientific experts failed to produce the desired authentication, their conclusions would be disparaged.

The Shroud of Turin Research Project of the Catholic Church began in 1978 with the aid of several clerics and a forensic chemist, Walter C. McCrone, renowned as the world's leading expert on art forgery. McCrone found traces of red ocher, vermilion, and rose madder—pigments that artists used in the Middle Ages to depict blood. He also found yellow ocher, orpiment, and azurite pigments along with ultramarine and titanium—materials most readily available from a medieval artist's studio. McCrone's secrecy agreement prevented him from publicly presenting his results, even while statements issued by the church misleadingly implied that no artist had painted on the shroud. About the same time a Swiss criminologist, Max Frei, claimed the shroud had 33 different pollens, all of which, others agreed, were from Palestine and the area around Constantinople. After Frei's death in 1983, it was determined that only one sample of the shroud had these Middle Eastern pollens, the other cloth samples had virtually no pollen grains. And none had pollen from olive trees that should have been in bloom at the time of the crucifixion. Given Frei's prior legal difficulties, one wonders whether one sample had been "accidentally" contaminated.

In order to settle uncertainty about the shroud's age, three laboratories set about in 1988 to use radiocarbon dating (carbon 14) with accelerator mass spectrometry on samples of thread taken from several regions of the shroud where there was no indication of threads from recent repairs. For their samples the three laboratories independently agreed on the date of the cloth as 1325±65 years CE, close to 1355 CE when it was first displayed in public. By 2001 ten published books had concluded the Shroud of Turin was woven in the 1300s, as against more than 400 books claiming the shroud either was, or could have been, the authentic winding sheet of Jesus woven more than a millennium earlier. The power and persistence of human desire and the Vatican's refusal to commit more samples of the cloth for analysis ensure that adoration will continue. Every summer in Turin, Italy, curious tourists mingle with throngs of devout believers to venerate the Shroud of Turin.

Recurring miracles could also be scientifically tested. For example, around Naples, Italy, the congealed blood of some martyred Catholic saints has a pronounced tendency to liquefy during religious ceremonies. Since appropriately prepared blood-like mixtures will liquefy during agitation or warming, perhaps the religious authorities might be interested in an objective analysis of these samples of "blood".

Religious Mythology Enters a World of Evidence

Because religion is invested with sustaining and propagating a settled belief system, new scientific findings may conflict with religious creeds. In modern times religion has given ground, sometimes grudgingly to scientific advances in understanding, especially in evolution and molecular genetics.

Some fundamentalists effectively advise us to "abandon all attempts to understand nature, ye who live here on Earth." For those whose conscience requires basing their actions on evidence, this kind of advice is as repellant as the famous inscription at the entrance to Hell in Dante's Divine Comedy, "Abandon all hope, ye who enter here." The dispute with fundamentalism is not about the importance of order and discipline and decency in life, it is about whether one seeks answers from faith or from evidence. When people don't have the answers required to structure their lives, they may make up some plausible answers since the human brain desires a world that makes sense. It is too disorienting to live in an unsettled swirl of inexplicable events. Fortunately, most events today have understandable material explanations, if we embrace the relevant evidence. Differences in the belief-sets of science and religion are a reflection of differing methods of inquiry and different domains of interest.

12

&

Scientific and Religious Quests for Truth

Seeking Truth through Science

To review briefly, scientists make discoveries by observing and testing properties of nature. Observations that cannot be repeated are not persuasive. Accurate predictions and practical applications help to confirm the truth of scientific claims. Since scientific conclusions are provisional, they may be modified by improved assessments. A supernatural intervention, or some other proposal that is untestable in principle, is not part of science. The methods of science and its requirements for proof are largely the same worldwide. Apart from a well-educated, inquiring mind, science erects no intentional social barriers to research and discovery. One can practice the same science regardless of ethnicity, gender, or nationality if one has solid training, some ingenuity, and suitable resources. These ideals sometimes falter in practice.

Religion and Doctrinal Truth

Validating religious power. Skepticism about religious power and doctrine has a long history at the highest secular and religious levels. In the mid-13th century Pope Innocent IV sent a letter to the grandson of Genghis Khan (Guyuk Khan) ordering the Mongols to reveal their future intentions and to cease assaulting Europe and persecuting Christians. The letter stated that God had delegated all earthly power to the Roman pope, who alone was authorized to speak for God. Responding in a letter, which still survives from 1246 CE, Guyuk Khan asked the pope how he knew whom God absolved and treated mercifully. After all had not God given the Mongols, rather than the pope, control of the world from the rising sun to the setting sun? Khan even invited the pope to visit and pay homage since God expected the Mongols to spread his commandments and his laws through Genghis Khan's Great Law (Weatherford, 2004, p. 163). This dispute is reminiscent of an earlier standoff in 1054 CE when the Roman pope and the Eastern Orthodox pope exchanged letters of excommunication.

Verifying holy scripture. The Christian Bible, the Hebrew Bible, and the Qur'an and its coupled commentaries are widely revered sources of religious belief. Many Christians believe the New Testament is accurate. Many Muslims believe the Qur'an is accurate. Members of the Church of Jesus Christ of Latter-day Saints believe that Joseph Smith's The Book of Mormon (1829) is accurate. Scientologists believe that L. Ron Hubbard's Dianetics (1950) is accurate. Which, if any, is really accurate? The Bible includes historical and empirical claims, moral advice, supernatural connections, and influences like miracles and messages from the divine, and additional supernatural features like souls, Heaven, and immortality. Which of these biblical features have been verified as accurate? What certifies the credibility of scriptural passages?

May we expect doctrinal revisions? Right from the start of Christianity, religious documents that challenged existing doctrines risked rejection and suppression. The Gnostic Gospels, like the Gospels of Thomas and of Mary, are good examples of texts that warranted public consideration, but which early Christian bishops suppressed (Pagels, 1979). Clerics in the Middle Ages were no more tolerant of differing views. For example, although the brilliant Michael Servetus (1509-1553) discovered the major blood vessels that interconnect the heart and lungs, the Catholic Church resoundingly rejected his heretical religious views. For some time he managed to evade the outraged Inquisitors. No less angry, the Calvinists seized Servetus while he was praying on his knees in their church. They proceeded to burn him at the stake with copies of his offensive religious writings tied to his arms. Even now, religious institutions are reluctant to update their beliefs since continual change implies vacillation that undercuts the certitude required for comfort. Driven by a desire for certainty and fearing an unraveling of the credibility of holy narratives, a religion may tightly embrace its perceived truths. After all, te tive claims and a suspect creed are unlikely to retain followers. The stability of religious doctrines helps to anchor lives during times of rapid technological and social change.

As long as a supernatural force remains hidden and unmeasurable, it relinquishes its capacity to explain anything, not least its existence. For example, does it clarify "immortality" or merely add more unknowns to describe blessed immortal Salvation as a perpetual life in a Heaven ruled by God? In what sense is it a clarifying explanation to wrap the enigma of immortality in the mystery of an unseen Heaven overseen by an invisible host called God? Yet, as the devout appreciate, to argue that theological conjectures do not qualify as verifiable explanations does little to wash away our need for the spiritual. Our hearts are going to be persuaded to believe in God and an afterlife by feelings that soar and swell within us. It is our feelings that move us to spiritual beliefs.

Establishing Truth: Science and Religion Compared

Scientific naturalists believe both that the mind is a description of brain states, not a free-floating entity installed independently of the brain, and also that no vital spirit is required to animate life. Haught (2007) sees the mind and life as manifestations of the vitalizing power of the Holy Spirit, as an expression of the divine. Theologians believe there is more in the universe than matter and energy. Religious postulates like the soul, resurrection, immortality, Heaven and Hell, Satan and angels, miracles and other supernatural factors cannot be directly addressed by scientific methods. Most scientists believe these are figments of human imagination, although they cannot disprove the existence of the supernatural.

We have arrived at this impasse because science and religion are strikingly different in how they attempt to certify truth. One of the most interesting outcomes is the resulting stark contrast between the near-unity of scientific beliefs and the great diversity of religious doctrines. This difference arises in part because decisive scientific experiments are generally able to exclude erroneous formulations. A one-thousand-page textbook of biochemistry, for example, has few scientific errors because it is based upon tens of thousands of experiments that established truth by ruling out wrong explanations that were tested and discarded. Adding well-established facts to the collection of scientific information is like a garden enlarged with new flowers at its margins that replace weeds of confusion. It is exhilarating to carry out a decisive scientific experiment that rules out a plausible alternative explanation. The elegance of some experimental designs is evident in their clever simplicity.

Box 12.1 Exploring the Basis of the Sense of Pain.

Suppose you want to know how it is that a hard pinch to the skin triggers pain. In particular, does pain arise from lots of activity in readily aroused sensory nerve fibers, or does pain require activity in a special set of pain sensory fibers that only respond to intense stimulation of the skin? Edgar Adrian settled this issue with an elegant experiment. Adrian used the equivalent of a soda straw to direct a steady stream of air onto a frog sitting on a table. By dimpling the skin the air pressure aroused many sensory fibers, yet the frog didn't hop away. Then Adrian interrupted the steady stream of air with the blades of a spinning fan so that the skin was stimulated at more than 100 puffs per minute. Even though thousands of touch fibers in the skin were responding vigorously, the frog still didn't budge in response to the air puffs. But with the slightest poke of a sharp pin the frog hopped away. Evidently, pain was not caused by the sheer quantity of sensory activity but by the selective activation of a special class of skin sensory fibers—the pain fibers. Since Adrian's time, hundreds of more elaborate experiments have confirmed that our skin has at least four different classes of sensory nerve fibers that mediate the senses of pain, warming, touch, and cooling. In the characteristic British reserve, Adrian was known as a "clever chap." Lord Adrian was awarded the Nobel Prize in 1932.

Myths flourish in the absence of credible methods for ascertaining truth about the supernatural. There is no generally accepted way to separate truth from fanciful accounts about the immaterial realm. In contrast to general agreement within science, one finds an absence of decisive resolutions of conflicting views across different religions. If the supernatural were more readily accessible, religions would be able to evaluate and jettison many bogus claims while feeling reassured about the facts that remained.

The Extraction of Religious Truth

The sources of religious truth include: tradition, experience, reason, revelation, and scripture. Similar to enlightenment in the tradition of eastern religions, Christian revelation is personally experiencing God, often accompanied by a feeling of affirmation.

The single most difficult question for proponents of a religious creed is, "How do you verify what you claim to know from revelation or other sources?" Where is your evidence or your accurate prediction? There is no direct way to evaluate claims for the divinity of Jesus, or the presence of original sin and its absence in Mary, the proclaimed virginal mother of Jesus. If a factual dispute flares up over the authenticity of a particular historical event or a venerated artifact, further historical sleuthing and forensic dating of objects may settle this technical dispute for those willing to heed the evidence. To establish the correctness of a scriptural assertion, it would be ideal to have independent substantiating verification, such as historical documentation. Assessments in the 18th and 19th centuries led to "higher criticisms" as historians and theologians revealed repeated human manipulation of biblical texts, as we have previously discussed. Just at the time the Bible was coming to be more critically evaluated, evidence for biological and geological evolution was independently creating disbelief in the content of some scriptural narratives. In writing about his concern over the impact of Darwinism on scripture, Charles Hodge (1874) said, "Religion has to fight for its life against a large class of scientific men." Today, Hodge would gasp at the torrent of biological discoveries that are flooding theological repositories.

The Advances of Science May Alter Religious Beliefs

Scientific findings sometimes illuminate religious viewpoints. The publication of Charles Darwin's *The Origin of Species* in 1859 confronted theologians with four piercing questions. Is the Genesis account of the origins of life accurate? Are humans a special creation apart from other animals? What message does one take from the apparent designs in nature? And how can natural selection and survival of the fittest be squared with traditional moral theology? Ian Barbour (2000) has compiled some of the diverse theological responses. The most conservative Protestants remained steadfast that Genesis was correct. Humans were a special creation, and nature's apparent designs were evidence for God (Figure 12.1). The American Fundamentalist movement that began in 1909 rejected evolution because it challenged biblical literalism and appeared to support atheism. Other traditionalists were more accepting of evolution because it seemed to hold out hope for progressive improvement that could be under the direction of God. For some modernists God became a part of nature rather than a force superimposed to control and direct evolution. Lyman Abbott was one who saw the action of God as an immanent force within nature. For Henry Ward Beecher evolution did not change the moral grandeur of man. As people sorted out their feelings, it was common for individuals within the same denomination to hold divergent opinions about evolution.

Figure 12.1 Man from the Hand of God. This immense sculpture captures the awe-struck viewpoint of many biblical literalists. Humans are properly above nature because they were specially created and lofted by the hand of God. God resembles a superhuman who is owed grateful worship by diminutive humans precariously dependent upon his support. Alternatively, evolutionary biologists see humans as arising by natural processes through a long line of descent from simpler organisms. This statue by Carl Milles is located at his seaside estate in Stockholm, Sweden.

The Anglican Church has accommodated diverse views ranging from the traditional to the contemporary. Rev. Aubrey Moore agreed with St Augustine that the soul was not divinely created but rather was part of bodily inheritance. He pointed out that those who thought God only occasionally interfered with nature would have to conclude that God is usually absent. In a stance that seemed to accommodate evolution, Frederick Temple, just before he became Archbishop of Canterbury, observed that God did not currently interfere, but "...made things make themselves." (By this he probably meant that at the outset of life God put in place the mechanisms that underlie evolution or descent with heritable modification.) The Catholic Church initially repudiated evolution, but over the course of a century took a more conciliatory stance, aided by the traditional flexibility that the church gives to Roman Catholic theologians in interpreting Genesis and other scriptural passages. While evolutionary theory succeeded in unifying broad expanses of observations in biology, it paradoxically managed to diversify viewpoints in religion. For the most part theologians seized upon the portions of evolutionary theory, which merged comfortably with their traditional outlook, while ignoring or dismissing much of the rest.

Can thoughtful efforts to develop a scientifically informed theology weave a doctrinal fabric without holes? Like Jesus, traditional Christians expect to be resurrected body and soul. A one-and-done scenario for human life brings a chill to those longing for a supernatural way to extend their lives. Science offers no breakthroughs on this point, but religion has some deeply appealing scenarios. Consider the following reflection of the depth of God's commitment as he empties himself to benefit mankind.

> The theme of the descent or humility of God is entailed by the christologogical and trinitarian teaching that God is one with the person and fate of Christ... However, the divine descent as Jesus in no way means that God is weak or powerless...[This] image of a self-emptying—and hence intimately *relational*—God, the absolute outpouring of goodness and love, is the very essence of Christian experience of revelation. (Haught, 2007, p. 42-43)

In other words after a fulfilling Christian revelation, we can have the benefits of God's limitless strength along with his intimate, humble sacrifice on our behalf in Jesus. More generally, during religious revelations one dwells in the ambience of God, and receives comforting reassurances, perhaps even messages or directives.

Contemporary Christianity: One Theologian's Effort to Harmonize Theology and Science

In trying to retain the traditional views desired by their flocks and expected by their superiors, many theologians shun accommodation with science. In contrast, the theologian John Haught has tried to frame Catholic beliefs to be compatible with scientific findings (Haught, 2007: *Christianity and Science*). In so doing he reaches out well beyond material naturalism, which is the view that there is nothing except matter and energy. Haught's focus is on revelation.

On God, the universe, science, and immortal life. God is inaccessible to science because God represents both the greatest mystery and a still unfinished future vastly exceeding our universe. Both the existence and the specific characteristics of the universe "are products of the free decision of the Creator" (Haught, p. 132). The role of science is to explore and describe these characteristics.

> As far as Christian theology is concerned, however, it is in no way contrary to scientific accounts of life's emergence to attribute the existence of life ultimately to the vitalizing power of the Holy Spirit. (Haught, p. 151)
>
> ...God, in self-effacing humility and fidelity, acts powerfully and effectively in the world without violating the laws of nature. Such is the God, we may add, who can also raise the dead to new life [by nature's laws?]. (Haught, p. 160)

God's most visible interaction with Earth was his descent as Jesus in whose suffering God demonstrated love and compassion for humans.

> Christians believe in bodily resurrection, and bodies are inseparable from the material universe [which science claims is slated to burn out and "die"]. In some sense, therefore, resurrection, if it is not an irrational belief, must be the destiny of the entire universe, not simply of perishable human lives. If Christ is not risen from the dead, then vain is our faith and naturalism wins. But if he is risen indeed, then theological consistency requires that we bring the entire universe into the sweep of what is destined for redemption. (Haught, p. 155)

Seeking truth through revelation. Religion relies on revelation to validate fundamental propositions out of reach of scientific investigation. Revelations, which contribute to perceived religious truth, may emerge as a surge of feeling or a sudden realization. Some theologians claim the very nature of revelation places it beyond verification by science. If religious experience proceeds through revelation or the unveiling of mystery, precisely what is it that is revealed and how do we know the revelation is accurate?

Haupt believes that if you first adopt a religious belief-set on faith, then revelations that increase the desire to know and experience the promise of the future mean the revelation is correct. In his own phrasing:

> What I propose, then, is that the trust that awakens in you when you allow yourself to be taken hold of by the revelatory image of God's descent and promise [in the form of Jesus] can function to liberate and promote the interests of your desire to know. If this claim turns out to be a reasonable one, then allowing yourself to be moved deeply by revelation will have satisfied the fundamental criterion of truth. (Haught, p. 182)

> Only after surrendering to the claims of revelation, and at the same time having become fully aware of your desire to know, are you in a position to assess the truth status of your faith. (Haught, p. 183)

So if you commit to the Christian belief-set and your revelation increases your desire to know, then you will engage truth. Isn't the truth more likely to emerge if we are not pre-committed to a belief-set? Or is it that God will only communicate with the precommitted? Since the desire to know is simply a starting point for empirical research, most scientists will find it quite puzzling that mere openness to truth will generate truth.

In recognizing that revelation is a manifestation of faith, Haught says,

> ...we would do well to call to mind once again that for Christians revelation comes in the form of a promise. So we are simply not in a position to verify or falsify it in a publicly comprehensible way at present. Only if and

> when a promise comes to fulfillment could our trust in it be fully vindicated. (Haught, p. 179)

In matters of morality should Christians rely on Natural Law and scripture? Is the cold pitiless hand of science unable to sift out moral rules from material enlightenment?

13

&

Some Religious and Secular Views of Morality

From Religion to Relativism—Five Diverse Views in the Competition for Human Values and Loyalties

Most **major religions** offer a supernatural power (God) and the prospect of an afterlife (Heaven or reincarnation). Non-religious **skeptics** dismiss the supernatural as fairytales that offer mysteries without clarification. **Communism** has dismissed God only to degenerate into personality cults (Lenin, Stalin, Mao, Castro, Kim Il-Jong, and others)—a kind of domestic demigod intent on autocratic rule. The clash between communism and Catholicism is a collision between two autocracies competing for the control of humans and their loyalty. Much like a theocracy, communism has no interest in sharing its monopoly on power. In contrast, **science** draws respect and power through its ability to generate understanding and practical applications to the natural world without coercing loyalty, although it has the potential for intellectual intimidation. Finally, **cultural relativism** holds that every society is entitled to its own values that are all valid because they are all sincere, as if sincerity somehow creates truth and erases conflicts among beliefs. From encounters with various religious and secular outlooks like these, each of us will develop a worldview that reflects our relationship to religion, political systems, science, and cultural values. How we come to our worldview and how we modify it says much about who we are.

Questions about Moral Advice in the Christian Bible

The Bible has both uplifting spiritual advice and harsh commands—the compassionate is intermingled with the cruel (Green, 1979). To have one's beloved children rely on a literal adherence to the Old Testament for their family values, should install more of a sense of absolute dread in grandparents than reassuring comfort. Would you advise your grown children to embrace any of the following Old Testament commands?

In the treatment of a disobedient child:

> If a man have a stubborn and rebellious son, which will not obey the voice of his father, or the voice of his mother, and that, when they have chastened him, will not hearken unto them: Then shall his father and his mother lay hold on him...and all the men of his city shall stone him with stones that he die. (*Deuteronomy* 21:18-21)

Elsewhere the Old Testament advises havoc and mayhem, and brutal killings, even of children.

> Behold the day of the Lord cometh, cruel both with wrath and fierce anger, to lay the land desolate: and he shall destroy the sinners thereof out of it...Everyone that is found shall be thrust through [stabbed to death]; and everyone that is joined unto them shall fall by the sword. Their children also shall be dashed to pieces before their eyes; their houses shall be spoiled, and their wives ravished. (*Isaiah* 13:9,15,16)

> And Moses said unto them, Have ye saved all the women alive? ...Now therefore kill every male among the little ones, and kill every woman that hath known man by lying with him. But all the women and girl children that have not known a man by lying with him, keep alive for yourselves. (*Numbers* 31:15-18).

> O Princes of the House of Israel, who pluck off their skin from off their bones: who also eat the flesh of my people, and flay their skin off them, and they break their bones and chop them in pieces, as for the pot, and as flesh within the cauldron. (*Micah* 3:1-3)

> Happy shall he be, that taketh and dasheth thy little ones against the stones. (*Psalms* 137:9)

With the possible exception of today's torturers hired by governments, it would seem certain that if you were to carry out such scriptural commands today, you would be jailed as a criminal. Nor does the New Testament offer uniformly comforting scripture. Consider the advice of Jesus when he says,

> Wherefore if thy hand or thy foot offend thee, cut them off and cast them from thee: it is better for thee to enter into life halt or maimed, rather than having two hands or two feet to be cast into everlasting fire. And if thine eye offend thee pluck it out. (*Matthew* 18:8,9)

It doesn't encourage family values for Jesus to have said,

> Think not that I am come to send peace on earth: I came not to send peace, but a sword. For I am come to set a man at variance against his father, and the daughter against her mother, and the daughter-in-law against her mother-in-law. And a man's foes shall be they of his own household. (*Matthew* 10:34-36)

It is imperative that we move beyond the harsh advice of ancient scripture.

Religious and Evolutionary Sources of Absolute Moral Rules

Religion can offer moral values arising from a mixture of scripture, tradition, and the instructions of historical prophets and living authorities. But is it the case that each religion provides valid or correct moral guidance? None of us follows scripture blindly, even those who are most enraptured with its inerrancy. We select pleasant scriptural quotations, giving them favorable interpretations. When we cherry-pick those verses we find attractive, we are selecting reflections of our own desires and judgments from the much wider range of values expressed in scripture. Are there alternative guides apart from constructing our own?

Absolutism leads us astray. It can be unnerving when an individual professes absolute values. Such a person may adopt absolutist views when their needs overwhelm personal honesty that would require them to admit they lack substantiating evidence. If people willingly deceive themselves by professing absolutes, isn't it likely they will be willing to deceive others in the moral arena? One of the greatest dangers to civil conversation, including peace between tribes and nations practicing different religions, is moral absolutism and the character of the individuals who have a need for it. "If you disagree with me, you are the one who is wrong." It is a short step from being judged wrong to being condemned as evil. You can count on the absolutely righteous to reject or cast aside or even attack as heretical and evil those who disagree with them.

For different reasons extreme absolutism and pure relativism share an abhorrence of being critiqued. It is hardly surprising that even a whiff of relative values annoys Pope Benedict XVI because cultural relativism threatens to undercut the papacy's aim of preeminence. Even while positioning itself as the citadel of absolute authority, the Vatican reveals defensiveness. Consider the Vatican's dismissal of Israel's criticism that Pope Benedict XVI deliberately omitted the name of Israel among those nations suffering suicide attacks: "The Holy See cannot take lessons or instructions from any other authority on the tone and content of its own statements," said the Vatican on July 29, 2005 in its haughty rejoinder to Israel's complaint.

Natural law. On what grounds does a theologian proclaim a particular "Natural Law" as absolute if it requires a human being to proclaim it as a true, divinely approved moral principle? When told that certain morals follow from Natural Law, we would like to know the grounds on which humans compiled this legal manual. Absolutists may appeal to Natural Law as a way to end the discussion, even though it is evident from the existence of deadly toadstools, festering intestinal parasites, rabid vampire bats, and poison ivy that not everything natural should be embraced. It is farcical for a priest sworn to celibacy to instruct married couples to obey Natural Law and have children. The reality is that the various sets of absolute Natural Laws are human inventions that may or may not have a natural basis in biological predispositions.

Settling Upon Simple Universals and Practical Guidance in Morality

We can begin by asking what rules humans should follow. The path to a general theory of morality is notably strewn with the schemes and carcasses of moral philosophers massacred by their critics. Compiling a long list of moral absolutes is not a path I propose to travel. Invented moral systems tend to be arbitrary and usually unworkable. Nonetheless, history offers ˎ gestions.

Fair play. Moral principles are statements about the way liˎ creatures should interact. We can expect to achieve a strong consensus for bedrock moral rules. Fairness is one value. Fairness is the foundation for the view that everyone should abide by the rules or in more general terms that we are all equal before the law guided by a set of rules and legal procedures that we would want for ourselves. Everyone wants to be treated fairly. C. S. Lewis (1942) argued that once we sense the Moral Law within us and admit that we have often broken it, we are primed then for rescue by Christian belief in a rule of fair play.

The Golden Rule. Beyond fair play, I offer sensible applications of the Golden Rule as a sound approach to human decency. The Golden Rule is hardly new, but in modern times one expects to have it more frequently followed. It says, "do unto others as you would have them do unto you" or in contemporary parlance "treat others the way you would like to be treated." To paraphrase Rabbi Hillel and others even before the time of Christ, "Follow the Golden Rule—all else is commentary." The Golden Rule works best when everyone shares the same value system. It may falter when there are substantial differences in lifestyles and values. You may need to allow others to have what they personally value, even if it's not your own preference.

The following behaviors fail the Golden Rule. We reject: that a thief should have his offending hand cut off, that vulnerable young women should be subjected to clitoridectomy to trim their pleasures or the likelihood of illicit sex, that married couples should be denied access to condoms which will prevent the spread of venereal disease and AIDS to infants, that women should be denied the right to vote, that young children should work 12-hour days in sweat shops, that it is acceptable to have captive or economic slaves or to torture the suspected or the guilty. If you personally want to be hauled off and subjected to any of these treatments, raise your hand. Otherwise please help to resist attempts to impose these conditions. We need to find a higher road that is humane and inclusive.

Postmodern relativism. In the unkempt stables of academe, postmodernists horse around with pretentious insights. They assert that the recent centuries of scientific enlightenment have generated only temporary and impermanent truths that differ across cultures. They say that scientific truths have no superior validity; they are just one of many equally valid social constructions, "one particular set of superstitions" in their postmodern phrasing (Gross and Levitt, 1997). Some postmodernists even claimed that clearly expressed ideas are tools of oppressors. It is certainly true that if we all adopted an intentionally muddled pattern of expression, then much less of anything would be accomplished, including oppression. Mercifully, the postmodernist movement, which had its greatest following in literary circles, has probably run its course. Postmodernism may nonetheless have some lasting benefits, such as a greater interest in the social context of judgments and greater awareness of potential conflicts of interest. In most human affairs, transparency of actions tends to promote more responsible behavior.

Punishment. Before we react vindictively against evildoers we should remember the longstanding rule in jurisprudence that the excessively harsh punishment of another's misbehavior is itself a crime. Consider the death penalty, which is illegal in most advanced democracies except for the United States. Killing by a State or by the United States government is intended to deter crime and to act as a just retribution in the sense of the Old Testament's, "eye for an eye." However, because most murders are crimes of passion or result from unanticipated errors, it is not surprising that capital punishment is an ineffective deterrent. For a decade the Innocence Project of the Benjamin Cardozo School of Law has used DNA to review convictions. DNA tests have identified nearly 200 wrongly convicted prisoners, with a similar number under review. Fourteen innocent individuals were on death row, scheduled to be executed—or rather, murdered by judicial misconduct. Justification for the death sentence plummets when the innocent are convicted. These 200 innocent prisoners had already been locked up for a collective total of 900 years before their recent release. The careers of ambitious prosecutors should not be built on the backs of the convicted innocent. At the outset, embittered crime victims who claim they can visually identify the perpetrator must receive pretrial instruction on the high rate of such misidentifications. Passionate for revenge, outraged victims may act with malice. It is hard to imagine a more irresponsible action than erroneous testimony that puts an innocent person to death. Officialdom is generally reluctant to study the problem of convictions of the innocent, but the power of DNA evidence is forcing them to correct at least some miscarriages of justice.

Science Can Contribute to the Identification of Moral Truths

Religion often claims dominion over matters of ethics and morality. That's understandable. However, the consideration of absolute morals and values is not outside the purview of science. Admittedly, at the outset the theory of evolution seemed to justify human selfishness. What is the point of being kind, if it wastes energy helping competitors at the expense of the survival of your genes? An organism's genes are an information system that perpetuated itself, often at the expense of other organisms. In spite of this bias, as social creatures, the odds of our survival are enhanced by the efforts and skills of friends and relations. Biologists say it is in our nature, in our evolved predispositions, that related humans co-operate.

Kin altruism and reciprocal altruism. When kin show mutual respect, it reduces abrasive interactions. Fair play is generally reciprocated. Fairness is embedded in the character of some other primates. Capuchin monkeys refuse to work if they see that other monkeys are being paid for doing no work or being paid more for the same work (Brosnan and de Waal, 2003).

By helping our kin who share our genes, we improve the odds of the survival of some of our own genes. Behavioral research backed by sophisticated mathematical analyses has shown that it can pay to be nice not just to your kids but also to your extended kin like nephews and nieces (Hamilton, 1964). After all, they also carry some of your genes. Remember successful genes are those that manage to be passed on to future generations. The survival of primitive humans increased when they embraced their kin and distrusted others. Because of its considerable survival value, the natural distrust of strangers, or xenophobia, is likely to remain a part of us. Such anciently rooted biases against strangers now threaten the stability of contemporary civilizations as small tribes gain powerful weapons to direct against distrusted foreigners.

Further studies of evolution point to positive natural selection for the behavioral exchange of favors as in "you scratch my back and I'll scratch yours." Such reciprocal altruism in the form of the mutual exchange of favors provides a selective advantage for your genes (Trivers, 1971). This is close to the Golden Rule of treating others as you want to be treated. Groups with highly selfish individuals probably lost out to groups with more co-operative individuals.

Through millennia of evolution each of us has genes that underlie some basic moral behaviors. We can conclude that kin altruism and reciprocal altruism are important moral values built into human behavior. Of course, there is ample debate about the character and strength of universal moral tendencies. Cultural variation suggests a big role for social training in the details of morality. Further, it is a big jump from having innate moral tendencies to actually adhering to a specific moral code. As we all know, it is one thing to receive moral training, it is quite another to follow it. Moreover, if evolution selected mainly for an interest in oneself and close kin, close adherence to broad community standards may not be a natural extension.

Morals and brain circuits. Neuroscience can be expected to inform us about the physiology of various moral and immoral behaviors and about the origins of some feelings related to morality. Biologists believe that evolution has generated brain circuitry favoring certain moral or social conventions (Gazzaniga, 2005). All thoughts and feelings, including religious thoughts and feelings, depend on brain activity. Oxytocin is a well-known hormone classically responsible both for the release of milk (lactation) and for muscle contractions during birthing labors (Kosfeld et al., 2005). This fascinating hormone arouses neural circuits that generate trusting feelings and follow-on behavior. Oxytocin may also enhance parental acceptance of the infant.

Disordered brain circuits may mediate the bouts of compulsive gambling that can occur in Parkinson's patients treated with excessive dopamine. Could a modest adjustment of dopamine neurotransmitters help to curb ruinous addictions to gambling in normal people? As it clarifies the neural bases of behavior, science may be able to quell some pathological immoralities.

In searching for absolute moral values we might seek out universal properties of our brains. We are certainly born with enough budding skills (Pinker, 2002), to squelch the argument that human brains are a blank slate lacking behavioral predispositions. The brains of human babies come extensively programmed for useful behaviors: crying, smiling, chortling, laughing, babbling, and imitating come easily. These built-in behaviors elicit nurturing feelings and protective actions in parents. Evidently, neither babies nor first time parents tackle child-parent interactions as blank slates.

Moreover, as highly evolved social creatures living in tribes, it is reasonable to suppose that humans have brain circuits that aid social bonding. All newborn humans are endowed with neural circuits that will control and adjust their behavior toward other creatures. Where there are universal needs for procreation and survival, one can anticipate a selective advantage for brain circuits that mediate those needs. Our genes construct us with remarkable fidelity to the instructions from a genetic code that is likely to build in some behavioral universals. It seems reasonable to assume that biological moral biases contributed to our survival. The capacity for guilt and contrition is a good example. A sense of guilt is an effective means of motivating moral behavior of those who live in groups. Does anyone claim that cats have a sense of guilt? The marvelous brains of cats make them remarkably independent. Trying to make a typical house cat feel shame is an unrewarding task.

Our evolved natures. Shared religious beliefs increase tribal co-operation in the tasks of providing shelter and food, and having to defend against intruders. Nonetheless, an evolved culture isn't the only determinant of our social affairs. Anthropologists say there are also individual behavioral universals like forming superstitions, engaging in rituals, establishing hierarchies, using standardized facial expressions, and negotiating with bargaining behaviors. Brown (1991) claims as many as 300 such behavioral traits are universals in human societies. Universal modes of behaving and reacting and valuing bring greater coherence to social groups. We draw resolve and meaning from our web of interpersonal connections and universal conventions.

Change is Looming on the Moral Landscape

In the near future the moral landscape will be transformed by several new developments in biology. At present there are major moral disagreements over fertilization, contraception, capital punishment, abortion, and aspects of stem cell research. Soon we can expect additional disputes over the selection or modification of embryos in order to eliminate or enhance particular traits. Many powerful religions are keen to establish their preferred moral systems in laws which regulate or even prevent these developments. Yet, if the demand is there, these medical options will be offered somewhere—available to those who can afford them.

In fertility clinics sperm are mixed together with an egg in a glass dish for in vitro fertilization. Couples who are known carriers of certain mutant genes may then screen their three-day-old embryos for possible genetic defects. Parents want to select for a healthy child and avoid implanting an embryo that carries a mutant gene, which will lead to an infant with a short and painful life. After one cell is removed for genetic analysis, the embryo rapidly adjusts by automatically replacing the removed cell, returning the embryo to the eight cells present on day three after conception. In perfecting such remarkable medical advances, genetic technology will make it more likely that at-risk couples carrying seriously flawed genes are able to select for healthy children.

Many parents look forward to nurturing a child who is both healthy and bright. Within the near future it may be possible to screen for components of traits like intelligence that depend upon multiple genes. Some people will properly ask where the "improvement" of developmental processes will end. People are not plants or dogs to be artificially bred. In contemplating marriage ought there to be testing for particular genes, in addition to the usual evaluation of one's partner based upon educational achievement and other social traits? Will we want to know a mate's genetic account along with their bank account? It is going to be difficult to prevent some privileged people in the next generation from genetically selecting for healthy and bright mates and children. What biological processes are best left to God and the vicissitudes of nature? This is a long and deep discussion, better done in clinical settings than by regulations set in legislative stone. We should certainly move beyond vague apprehensions to clarify how a genetic procedure will translate into specific benefit or harm.

Science is Laying the Groundwork for the More Humane Treatment of Animals

The tolerant attitude of Europeans toward stem cell therapy suggests some Americans are needlessly nervous about the corrosive effects of the theory of evolution on the exalted status of humans (see Chapter 14). After all, evolutionary concepts have converted nagging uncertainties and ignorance about the status of humans into a new appreciation of our deep roots. By revealing genetic similarities of animals and humans, molecular biology has dramatically raised the stature of animals. Their humane treatment is an inescapable obligation. The rights of animals will be a moral concern of increasing importance. If our wanton killing and violence against harmless birds and mammals subsides, it will mean that humans have become more humane and less engaged in blood sports that tend to devalue life. As we increasingly value animal lives generally, we will place a higher value on human life specifically.

Medical realities cause most people to assign human lives the highest priority. The reason we are comfortable with the rule that higher animals have a right to kill lower animals, is because we think, with no extraterrestrials in sight, we are the highest animal around. Even so, if we were visited by clearly superior extraterrestrials who wanted to use us for medical research to cure a particular illness, we would be strongly motivated to argue it is wrong to kill us to advance their health. So we must admit it, the animal activists have a valid point. But can we expect them to be true to their own values? How likely is it that a grievously ill animal rights activist will insist a hospital shall use no procedures from the extraordinarily long list of medical treatments, like vaccines, that arose from animal research? Those who reject euthanizing any mammal for the purposes of medical research risk being hypocritical recipients of medical services. And if a determined animal rights activist persisted in waiving off such treatments for himself, could we expect a conscientious doctor or nurse simply to let this person die when there is help at hand?

Morality and ethics deal with rules governing our direct and indirect interactions with humans and other creatures of feeling. Many have yearned for some absolute moral standard to guide their journey through life—preferably a code that provides a reliable autopilot. Parents generally guide their children so they cause no serious harm or evil. It would have been nice had God done the same for his human creations. He might have given our eyes yellow spots whose timely flashing warns us when a contemplated action might jeopardize our chances of reaching Heaven. If only it were so easy to receive moral guidance from God. A better understanding of evolution may help us appreciate our moral roots.

14

&

Coming to Grips with Evolution

Life is Improbable

It must be extremely rare for a chemical mixture to succeed in replicating itself with some heritable variation, which is to say it is alive! Even after life began, it took many millions of years for organisms to diversify into varied species exquisitely adapted to particular environmental niches. As Charles Darwin realized, it is much easier to describe the diverse adaptations of living species than to identify the selective pressures and changes that led to a given species arising over a span of thousands of years. Since favorable mutations are quite rare, the importance of long periods of time is crucial to understanding speciation. When stretched across tens of thousands of years the sporadic accumulation of favorable mutations allows natural selection to generate sophisticated species. Major multi-gene changes may have spanned millions of years.

The human species, *Homo sapiens*, is a relative newcomer. We have been around for only some 100,000 years. Although this might seem like a long time, we are fresh on the scene considering how long life has existed on Earth. If we imagine that the first unicellular life began 24 hours ago, then on that timescale modern humans sprang up less than 3 seconds ago!

In traditional theology, the clever designs of organisms implied an impressive supernatural Designer. Instead, Darwin saw complexity arising not from a Designer's hand but from the accumulation of adaptive changes that led from simple forms to increasing biological complexity through untold generations. Some envision this evolutionary development of complexity as blind and mechanical, a trial and error process. Less starkly, the ancestors of successful organisms evidently took advantage of their opportunities. Those forms of life that benefited from their particular heritable variation were more likely to survive and multiply. There is no evidence that the evolution of biological complexity was planned or guided by some thought process, nor is there any need to postulate such a process to account for the grandeur of nature as we know it.

In the Grasp of Apes

Darwin (1809-1882) was among the first to shift from the theological view of permanent perfection to the reality of ceaseless organismal change. One riveting tale recounts a testy verbal exchange between Bishop Wilberforce and Thomas Huxley at a gathering at Oxford University in 1860. In this debate over Darwin's *The Origin of Species*, Wilberforce asked Huxley, "Is it on your grandfather's or your grandmother's side that you claim descent from a monkey?" Huxley deftly replied, "I would rather be descended from a monkey than from a man who abuses his great gifts to distort the truth." It is not hard to imagine a red-faced bishop angrily defending the God-given dignity of man. However plausible this sharp exchange between Wilberforce and Huxley may appear, it probably never occurred. It appears to be an apocryphal tale embellished by the press nearly 40 years after the event. As a professor of theology and mathematics at Oxford, Wilberforce was by other accounts a reserved gentleman. Indeed, it is said that Wilberforce's measured critique of evolution at this meeting in 1860 sufficiently impressed Darwin that he went back to his study to sharpen his arguments and analysis. That is how an exchange of views in science ought to proceed.

So how does it stand now for exalted humans vis-à-vis monkeys and apes? It doesn't do much for our pride to think that humans are dangling from a twig of one branch of primate evolution. You don't need much of an ego to resist the humbling conclusion that "except for a few stretches of DNA that were lost or gained or rearranged, there would go I, just a knuckle-walking, hairy ape." The decoding of the human genome in 2001 by the International Human Genome Sequencing Consortium (Lander et al., 2001) and the initial sequencing of the chimpanzee genome in September of 2005 by Mikkelsen et al. make it undeniably evident that humans and apes are closely similar. How similar? Consider the

3.3 billion bases available in each genome (every cell's nuclear DNA contains one copy of the complete genome). Now suppose that we randomly pick a short sequence of 85 bases from a corresponding stretch of human and chimpanzee DNA. To make the similarity readily evident let's convert these bases into 85 English characters in this poetic form:

> Breathes there the man with soul so dead,
> Who to himself hath never said,
> This is my own, my native **land**!
> (Walter Scott 1771-1832)

> Breathes there the man with soul so dead,
> Who to himself hath never said,
> This is my own, my native **band**!
> (Student version)

Would you accept the student's argument that in his version the one-letter difference in these stanzas of code meant he didn't need to cite Walter Scott as the source, which he couldn't remember at the moment? "After all," the student continued, "my single letter substitution changed the poem's meaning from a patriotic statement to a musical metaphor." But just as with genes, even though we agree that a slight modification may significantly alter function, the common root is plainly evident.

In looking at millions upon millions of letters of DNA code it is evident that humans and chimpanzees differ at about the same percentage level as these two versions of Walter Scott's stanza (we can neglect copy-number variation—the major deletions or duplications of DNA). A variety of evidence indicates that humans and chimps shared a common ancestor that roamed the forests of Africa more than five million years ago. Just as predicted from earlier research on primate evolution, studies of human and chimpanzee genomes show striking similarities. If we ignore individual variation, about 99% of the DNA bases in human and chimpanzee genomes are identical. This still leaves plenty of room for species differences, since a 1% difference is equal to some 33 million bases. The occasional differences in the DNA code include some changes that are quite important to humans. The differences are the greatest in the Y chromosome, which is unique to males. Of the roughly 25,000 functional genes in humans, some 11,546 differ, however slightly, from those in chimps. We can expect that some stretches of DNA, which make us distinctively human, will soon be pinpointed. I hope we will all be proud of these altered proteins and the added-in and dropped-out traits.

Changing the DNA Code

Organisms can't biologically evolve unless they acquire some genetic changes that can be inherited. Rearrangements and other mutations in DNA can produce heritable alterations in the genetic code. Even though the sequencing of organismal genomes is providing a new and powerful outlook on the history of changes in the genetic code, it remains difficult to reconstruct mutational histories over millions of years. What about recent changes? Is there evidence that mutations are presently occurring in some organisms? Most certainly. Some of these mutations seriously threaten human populations, like mutations in influenza viruses. There are dozens of mutant forms of the flu virus. Public health officials are desperately concerned that ongoing mutations in the avian flu virus will produce a mutant form able to spread readily from human to human with no vaccine available to stop it. More than 300 million people might die, considering that 50 million died during the Spanish flu pandemic of 1918 when the world was not as populated and the spread of viruses by international travel less common. Compared to flu or other viral DNA, human DNA mutates at a slower pace, perhaps one change per fertilization. How do changes occur in the DNA code for humans?

Mutations. A piece of a chromosome may be duplicated, flipped end for end, deleted, inserted into another chromosome, or exchanged with a piece from another chromosome. It is not unusual for a mutational change to insert a duplicate copy of a sequence—often as long as several thousand nucleotides. When a chromosomal rearrangement splits the middle of a gene, it can have dramatic consequences.

A point mutation is the smallest change that can occur. It is the gain or loss of one nucleotide, or the exchange of one nucleotide for another, like an adenine replacing a guanine. Because of the high precision of DNA duplication and the presence of several error correction processes, point mutations are rare—perhaps one in 10 billion bases. So with 3 billion pairs of bases in the genome, some matings will have no point mutations. To understand the potential impact of the rare point mutation we need to appreciate that nucleotide sequences are read-out three letters at a time—the three-letter reading frame. The gain or loss of a single nucleotide might seem relatively harmless, but it can be devastating if it causes a "frame shift." Imagine reading this DNA code three letters at a time: THECATATETHEHAT. With blank spaces added for clarity, it resembles a five word English sentence: THE CAT ATE THE HAT. Now suppose that of these 15 letters (nucleotides), the letter C is deleted. Reading three letters at a time from the remaining 14 generates an unintelligible sentence: THE ATA TET HEH AT. This kind of frame shift can cause a gene to code for an incorrect series of several amino acids.

Conserved Species

It is not as though all organisms change form and function at the same pace. Sharks have retained the same streamlined shape for more than 425 million years. The king crab off the coast of Alaska is well-adapted to the stable environment of the ocean floor where it has little competition. King crabs look like their fossils from 100 million years ago. The same can be said for the ginkgo tree and the dawn redwood tree from China—"living fossils" as relics from the past. A species that earns the title "living fossil" is a true evolutionary success. Just as these species were admirably adapted to their environments, a gene, or more exactly its protein product(s), is well-adapted to its functional role.

Conserved Molecules

Every protein that functions as an enzyme has an active site where a target molecule binds to it. The exact shape of the active site and associated positive and negative charges are critical for the enzyme's action. In most cases the enzyme will no longer function properly if there is a substitution of one amino acid for another in the active site. In the outlying regions a different amino acid here or there may have little effect on the enzyme activity.

Cytochrome C. Harnessing energy for our cells requires a series of enzymes that includes cytochrome C. This is a highly conserved protein in the sense that its active site is nearly identical across more than 100 diverse species. For example, of cytochrome C's 104 amino acids, 33 are exactly the same in yeast, gray whales, sunflowers, sharks, wheat, fruit flies, turtles, Pekin ducks, gray kangaroos, rice, tunas, donkeys, Rhesus monkeys, *Neurospora* mold, humans, and many other species. *Neurospora* arose more than a billion years before humans, yet throughout this prolonged period, the active site of cytochrome C has remained unchanged. So much for the bogus idea that chance events riddle biology with instability. Would a human-made loom still weave the same patterned woolen sweater after millions of years of service? One can get persuasive visual evidence of the consistency of cytochrome C across organisms by comparing the remarkable similarity of the three-dimensional twists in the cytochrome C of edible rice and tuna fish (Figure 14.1). This ribbon of more than 100 amino acids is twisted and folded, and looped about in a beautiful complex 3-D pattern that is nearly invariant in rice and tuna. It was more than a billion years ago that the line of organisms that would give rise to rice plants split from the line that would give rise to tuna fish. Evidently, for hundreds of millions of years cytochrome C has maintained a remarkably consistent form across hundreds of species. During this period there must have been occasional mutations that inserted novel amino acids at the active site. We can assume these changes failed to improve function of the active site because these alternative forms are nowhere to be found today.

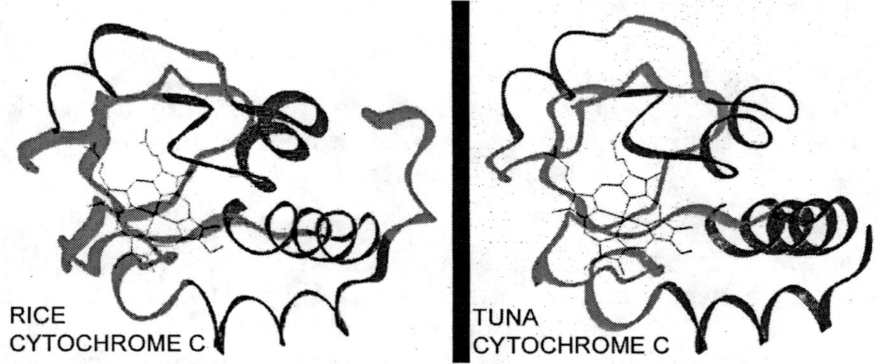

Figure 14.1 Three-Dimensional View of Cytochrome C in Rice Plants and Tuna Fish. The complex pattern of protein folding is crucial for protein function. The faint hexagonal array is an associated molecule containing iron. (Adapted from Purves et al., 1995 and others.)

Actin. For billions of years the protein actin has functioned in cellular movements of various kinds, including muscle contraction. Because it has been about three billion years since humans and yeast shared a common ancestor, it is hardly surprising that yeast has no physical resemblance to humans. In spite of that stupendous time span, human actin and yeast actin are so still similar they are interchangeable in mediating the movements of yeast cells and human muscle cells.

Insulin. Some 50 years ago scientists figured out the sequence of amino acids for a small protein known as insulin. As one of the vital hormones that regulate our blood sugar, insulin is in short supply in many diabetics. Following on from basic research in the 1920s, insulin-deficient diabetics are now able to inject an insulin solution, swallow it as a pill, or spray supplementary insulin into their noses to help regulate their blood glucose concentrations. The insulin protein is made of two chains: one of 21 amino acids and another of 30 amino acids. Let's examine the linear sequence of 21 amino acids in the so-called alpha chain of insulin. For convenience biologists give each naturally occurring amino acid a one letter abbreviation or code; G for glycine, L for leucine, and so forth. Mammals that look different may have similar forms of insulin. Here is the sequence of the 21 amino acids that make up the alpha chain of insulin (I have added some spaces to make some English-like words, just so the sequence is easier to read). Do you see any difference between human and pig alpha insulin? No, there is no difference. Further inspection reveals that rat alpha insulin differs from human and pig only in the fourth amino acid, D vs. E. The last row shows a random draw of the same 21 letters. As you might have expected, the odds were extremely low that a random draw of these 21 letters would give a sequence even remotely similar to alpha insulin.

Human:	GIVE QCCT SIC SLY Q LENY CN
Pig:	GIVE QCCT SIC SLY Q LENY CN
Rat:	GIV<u>D</u> QCCT SIC SLY Q LENY CN
Random:	LCVQ ECNC IGN YLY I SEQC TS

Among the 30 amino acids comprising the beta chain of insulin (which is not shown) humans differ from pig by just one amino acid and differ from rat by three amino acids. Not only do rats and pigs, and other mammals all use insulin to limit blood sugar concentration, but also as the analysis of the sequence of amino acids comprising insulin shows, these insulin molecules are closely similar. We can conclude that insulin is a conserved protein. That is, it performs its function so well, that it has changed little during the millions of years of mammalian evolution. Were a mutation to occur today, we would expect that the mutant insulin would not work as well, if it worked at all. An individual with a mutant insulin gene might be too sickly to thrive.

Histone 4. Mice and men look different as a result of differences in the regulation of gene expression that adjusts body proportions, the size and number of teeth, the distribution of hair, and so forth. Even though neither you nor I feel much like cozying up to a pig to say, "Hi there, distant cousin," we must acknowledge our remarkably similar molecules. As with numerous other proteins, the close similarity of insulin's amino acid sequences across mammals reveals their common evolutionary origin. Astonishingly, the histone 4 protein (one of a set of highly conserved histone proteins that are wrapped by DNA) has the identical sequence of 104 amino acids not only in mice and men but across a wide variety of species going back two billion years through deep molecular connections.

In contrast to the stability of histone genes, there are rapid adaptive changes in genes concerned with odor detection or immune recognition of invading microbes. Between these extremes of stability and rapid change there are many proteins whose amino acid composition has changed moderately over many thousands of years.

Most biologists believe that the reason today's mammals have such detailed molecular similarities as in insulin is that they all are descended from the same root stock—probably from rodents which managed to evade marauding dinosaurs. Another alternative, which is the hope of some, is that rather than evolution, it was God who in one stroke created man and other mammals. What no one disputes is that the construction plans of mice and men are remarkably similar. And comparisons between the sequencing of human and chimpanzee genomes show even more similarity; there isn't much about humans that is unique. Humans have arisen from occasional, yet telling, modifications of the stock construction plans for mammals. The beginning of the 21st century would seem to be an excellent time to follow wise theological instruction and not be prideful about who and what we are. It will not be long before we understand which specific genetic differences make us human and how some of these genes and their controlling sequences act. From detailed comparisons of the human genome with genomes of other primates we will come to learn which genes we have gained, which we have lost, and how changes in gene regulation led to the vaunted features that make humans admirable, at least in our impartial view.

Box 14.1 Tolerance of Lactose in Milk: The Co-Evolution of Human Genes and Cultures.

Secretory cells in the small intestine of human infants make the enzyme lactase that readily digests milk sugar (lactose). Lactase gene expression persists in most adult Northern

Europeans but in only 1% of adult Chinese. Many adults in southern Europe and Africa are also lactose intolerant because of low lactase expression. What might explain the genetic spread of sustained lactase expression?

About 7,000-9,000 years ago in Egypt and the Middle East the domestication of cattle yielded a reliable source of meat and milk. It was only some 3,300-4,500 years ago that cattle domestication as a way of life spread south of the Sahara desert into Kenya. In both instances this cultural change in dietary habits (drinking milk from cattle) was associated with an increase in adult lactose tolerance. Considering the calories, nutrients, and calcium in milk, the persisting ability of adolescent and adult humans to digest available milk sugar probably conveyed significant advantages in nutrition and survival. Recent genetic data suggests that beneficial mutations for increased lactase synthesis already existed, and so could spread quickly among humans who milked domesticated cattle. Specifically, within a stretch of a few hundred nucleotides on human chromosome 2, scientists have identified four single-nucleotide changes that promote lactase gene expression. These first of these mutations occurred 8,000-9,000 years ago in northern Europe. Some 7,000 years ago, three other point mutations occurred among pastoral East Africans. The study of Tishkoff et al. (2006) represents some of the strongest evidence for the dispersion by natural selection of recent favorable mutations existing in the human genome. Existing genetic instructions and new mutations are both subject to testing through natural selection.

Is Cloning an Undesirable Advance?

Cloning is a form of asexual reproduction like the rooting of a twig in potting soil. Cloning oneself means reverting to the more primitive asexual reproductive strategy which creates progeny genetically identical to the parent. Sexual reproduction is not useful for cloning because the mother's and father's chromosomes will be shuffled in each generation. However, one benefit of sexual mating is that progeny will gain some molecular novelty that thwarts the abundant parasites which have grown accustomed to your body's defenses.

In case you are spooked by the word "clone," remember that identical twins are naturally occurring clones; after all that is why we call them "identical." Yet, as we all know, identical twins are not identical in all traits, including their behavior. To find out how much your future clone might differ from you, we need to look at studies of identical twins raised apart. One finds that the shared physical and behavioral traits of identical twins don't guarantee identical lifestyles or outlooks on life. They are unique individuals with some different controls of gene readout and quite different memories reflecting their separate experiences in life. Nurture is also a major determinant of what we become.

Egoists tend to be flattered by the notion of cloning themselves. Regrettably for them, it is unreliable and unsafe to clone by replacing the DNA from a host egg with donor DNA derived from one of your body cells—a skin cell for example. The attempt is much more likely to produce a developmental disaster than a healthy replica of yourself. Similar problems may arise if a human skin cell is directly reprogrammed into an embryonic state and then encouraged to develop. Even if legal, the cloning of humans will be rare because of its great expense, and more importantly, because for every "success" there would be multiple failures resulting in deformed fetuses. Moreover, "immortality" through this route lasts only one generation. True genetic immortality would require repeatedly making a clone of a clone of a clone, appreciating that each replica would have its own personality. In any event the motive to repeatedly clone one specific genome is not likely to be sustained over multiple lifespans, let alone centuries. Regardless of these severe limitations, if Mozart or Elvis Presley were alive today, someone would surely entertain thoughts of cloning.

An alternative to consider is the bit of immortality available to us in the form of increased longevity. Although living longer sounds great, it could be troublesome if you are not healthy. And, in a process some say has started, one can imagine hordes of decrepit centenarians using their political clout to siphon resources away from the young in order to subsidize their vast ongoing medical expenses for drugs and treatments, including replacement parts from tooth to toe. And remember the adage, "I want to live as long as possible as well as possible." So will there be vim and vigor past 100? Welcome "Brave Old World."

How does human DNA stack up?

Even though our linguistic capabilities and some aspects of our intelligence are superior, human DNA has fewer genes than many "lowly" species. A salamander cell has easily 10 times as much DNA as in a human cell, whereas a simple garden lily cell has 20 times as much DNA. No less surprising, only 1.5% of our

DNA represents known genes that code for enzymes or structural protein. But scattered in the long stretches of DNA between the genes are codes for molecules that regulate gene expression. For example, much of our DNA produces RNAs of diverse functions. Even so, perhaps half of our DNA may have no function. Our DNA is contaminated by long stretches of DNA that duplicate themselves but don't seem to do anything useful—just boring sequences that repeat and repeat. We also have pseudogenes, remnants that have ceased to function. The ancient origins of some DNA are evident from some 50 genes that are apparently derived from bacteria whose DNA long ago merged with the DNA of our ancient single-cell ancestors. Similarly, ancient retroviral invaders translated their RNA into DNA that still contaminates our chromosomes. When the disorderly and dysfunctional sequences of human DNA were recently revealed, biologists became increasingly persuaded that we are part of a long line of descent, indelibly marked by parts and fragments of invasive microbes. Our DNA is far from the neatly arranged and carefully crafted DNA sequences an orderly designer would have produced. The human genome is a hodgepodge of make-do sequences littered with discarded relics.

The Impact of the Theory of Evolution on Human Belief

Like two varieties of cherry trees, two closely related organisms living in the same region will share a recent common ancestor responsible for the close similarity of their DNA. Darwin believed that all organisms are ultimately related. If you go back far enough in time, say two billion years, you could find the common ancestor of plants and animals. A slow but inexorable branching out of new organisms occurs when isolated groups of individuals mutate and improve their environmental fit. Living organisms are products of descent with heritable modifications.

Both the real and the imagined implications of this biological evolution trouble some people. If humans have evolved in a long descent from an unbroken line of billions of years of primitive organisms, then are humans just another type of animal? And might humans be neither a stable pinnacle of evolution nor a preordained creation of God? Moreover, if it was incredibly chancy whether humans would evolve at all, then does this imply that God never had a special plan for us, such as an intention to provide us with an afterlife in Heaven? In confronting these problems Pope John Paul II concluded that God allowed evolution to create a long line of descent of animals; at the end he installs each new human with a mind and soul. He believed that Catholics have little to fear from the biological descent of man because God was monitoring the process and looking after the spiritual realm, which includes the natural laws that govern morality on Earth and the immortality of the soul in Heaven.

Sensing their beliefs are at risk, religious fundamentalists raise two important questions about the evidence for biological evolution. "Don't the major gaps in the fossil record argue against continuous evolution where each species has closely similar ancestors and descendants?" Biologists reply that the apparent gaps reflect the rarity of fossil formation and the rarity of their discovery. One might have thought fossils would be abundant since perhaps 99% of all species that have ever lived are extinct. Sadly, because soft-bodied species don't fossilize and only a few bony or hard-shelled individuals were preserved well enough to leave fossil remains, there will never be complete closure on the step-by-step progression of speciation. Indeed, even at the most promising sites, discovering a novel fossil takes considerable time and a measure of luck. Nevertheless, in spite of the scarcity of fossils many of the major fossil gaps are filling in. Recently found major transitional fossils or "missing links" include "fish" with legs, amphibians with both gills and lungs, and whales with legs. From such new discoveries alone, it is obvious that the current absence of a particular fossil doesn't mean it won't be uncovered at some point. "The absence of proof is not proof of absence." (For recent fossil discoveries that help to trace the long-sought transition from fins to legs, see Figure 15.1.)

The second critique asks, "Where is an example of macroevolution or speciation?" Microevolution often reflects the enhanced expression of already existing genes. In a well-known example, finches with heavy bills begin to predominate when food is restricted to thick-walled seeds. We may think of macroevolution as the novel accumulation of major or multiple mutations leading to new species. Macroevolution frequently involves geographic isolation since this allows the accumulation of significant mutational changes because a breeding population has become separated from the main stock. In an emergent species the build-up of new mutations prevents breeding with the original population from which it came. The new species may not appear very different; it just won't crossbreed with the founder species. Species continue to diverge and become more distinct as mutations accumulate over thousands of years.

In contrast to the slowly accumulating consequences of living in separate habitats, chromosomal duplication provides a more rapid route to speciation. An occasional embryo may get twice the usual number of chromosomes because after a parental cell underwent the standard duplication of its chromosomes, the usual follow-up cell division failed to occur. For example, a century ago Hugo De Vries' plant-breeding experiments produced a primrose with twice the normal number of chromosomes. *Oenothera gigas* was a new species because it could not reproduce with normal primroses. Special types of wheat, cotton, and potatoes are examples of useful plant species that have arisen by chromosomal duplication. Similarly rapid speciation events are known for animals. The macroevolutionary formation of new species is occurring every day.

Religious fundamentalists are concerned that the widespread acceptance of the robust evidence for evolution would unweave the fabric of society and usher in moral decay and lascivious behavior. In addition to making humans feel we are no longer superspecial, evolution is seen as a threat to family. This may be the deepest fear of all. Fundamentalists fear their children will cease daily devotions and mealtime prayers of thanks, even leaving the fold of believers to become apostates headed for Hell. It would be emotionally wrenching to have one's children reject fundamentalist parental values and embrace the secular world. The lack of success in getting fundamentalist religious creeds to replace evolution in public school science classes (see Chapter 15) motivates some fundamentalists to shield their children from the impact of knowledge by opting for home schooling or by using taxpayer-funded parochial school vouchers. Lives trapped in dodging the evidence.

The Theory of Evolution is Poisonous?

By revealing our genetic connections with the natural world, science should be reducing our sense of isolation and alienation. Yet, among many religious conservatives, the word "evolution" has vile and reprehensible implications; it is a threat of demonic proportions. As evolution is repeatedly reviled in sermons, children are taught to abhor the word and the conclusion that we are descended from apes. We need to examine these feelings since it is unhealthy to ingrain innocent children with fear and anger. Demonizing evolution is unwarranted and damaging; it should cease. The hostility toward evolution is partially rooted in misinterpretations of Genesis and confusions about evolution. We ought to find accommodation in the correct understanding of evolution. Traditional time lines place Genesis in conflict with both geological and biological evolution. Scientists know with great certainty that the earth is billions of years old and that over much of this time single-cell organisms proliferated, eventually giving rise to a wide diversity of multicellular plants and animals. The present abundance of life on Earth has genetically traceable ancestral histories that extend back millions even billions of years. Conserved proteins like cytochrome C and histone 4 are indelible molecular signatures that link a series of organisms in long lines of descent.

But to understand about our descent requires a crucial piece of clarification. It is completely inaccurate, but all too often feared, that humans are descended from chimpanzees. Not so. Humans did not descend from chimpanzees like those in zoos today; the great apes evolved in separate parallel lineages. Here is what happened. Some six million years ago a single lineage of apes split into two branches; one eventually gave rise to humans and the other to chimpanzees. As a result of this split, our human line has been evolving for six million years on its own merry way largely independently of the chimpanzee line. That is a lot of time to evolve some stately characteristics. If a few decades are sufficient to erase the stain of a grandfather who failed to repay a loan on time, what kind of self-doting hubris is it that won't allow six million years of separate human evolution to erase the "stain" of an ape-tainted heritage? Such puffed up pride. Are we also to spurn our own deep ancestors, single cells from billions of years ago, because we fear being tainted by their frolicking, amoral antics?

The close relatedness of humans and chimpanzees is evident in the detailed similarity of their genomes. But more than four million years of accumulated genetic differences also matter; they are comfortingly numerous. The three million base pairs that differ are more than enough to account genetically for the profound differences in appearance and intelligence. The accuracy of these conclusions will be amply confirmed in the next decade or so. For example, after splitting off from other apes, the human lineage has undergone significant evolutionary change. Over 700 human genes are known to have evolved within the past five to ten thousand years alone, including several which affect brain development and function (Voight et al., 2006; Pollard et al., 2006). Genes that have been modified include some that control bone development, fertility, and reproduction. Others are involved in the processing of sugars, fats, and odors, such as food odors. Some genetic changes, like the spread of lactase promoter mutations, reflect the shift of humans to an agricultural lifestyle. Each genetic change is a story of utility tested by environmental pressures. All of this is good news for those who would like to distance themselves from chimpanzees. A bit of "bad" news is that the human and chimpanzee lineages probably got back together for some crucial cross-fertilizations before they separated entirely some four to five million years ago (Patterson et al., 2006).

Box 14.2 Finding Humanity in Sea Urchins and Neanderthals.

Ancestors of the lowly sea urchin are at the base of the massive deuterostome lineage that over millions of years gave rise to fish, amphibians, reptiles, birds, and mammals. One of the most startling revelations in molecular genetics is that sea urchins not only have about as many genes as humans but they include many of the same genes (Samanta et al., 2006). It had been commonly imagined that a passel of new genes accounts for the advanced properties of humans and other mammals. Evidently, the number and diversity of genes is not a particularly useful guide for distinguishing advanced recent organisms from primitive ancient ones. More than the emergence of new types of genes, the key to evolutionary change may be acquiring new ways to regulate the expression of old genes. To determine whether gene regulation sets humans apart from other creatures, we need to examine our closest relatives—living and dead.

Although chimpanzees are our nearest living primate relatives, their ancestors branched off from the human lineage some 6.5 million years ago. The human-like primates most closely related to us are the Neanderthals who died out only 30,000 years ago. They may have been exterminated by modern humans, who began to migrate from Africa into Europe some 40,000-50,000 years ago. Recent technical advances have made it possible to extract residual DNA from Neanderthal leg bones and decode one million base pairs, or about 0.1% of the Neanderthal genome (Green et al., 2006). Within two years it may be possible to decode the entire Neanderthal genome by continuing to analyze the DNA from one 38,000 year-old bone. Because the Neanderthals diverged from the human lineage less than 0.5 million years ago, comparison with their DNA will help identify some of the more recent genetic changes that make humans distinctive. Human and Neanderthal genomes seem to be 99.5% identical, as expected from their relatively recent divergence. Some interbreeding may have occurred after the divergence of the human and Neanderthal lineages, but the paleontological and genetic evidence is inconclusive. Like northern Europeans generally, the fair skin (with red hair!) of Neanderthals was probably an adaptation to boost vitamin D production during the winter's darkness (Lalueza-Fox et al., 2007). To understand the evolution of our primate kin is to increase one's understanding and respect for human life.

Fearing the Role of Chance in our Lives

Another distorted view of the evolutionary process that is emotionally wrenching is the claim that the beautiful forms and lives we see on Earth are entirely due to random events. This is not so. While chance is important, it is also constrained. Nor does chance automatically dissolve value or meaning. Preferring to feel we are decisively in control, we vastly underestimate the contributions of chance in our own lives. Chance in human affairs is responsible for new friendships and enjoyable spontaneity we wouldn't want to miss. Yet, the thought that our life in particular, and human lives more generally, are chance affairs, leaves some folks with a hollow, empty feeling, as if in granting a founding role to chance somehow makes life evaporate into meaninglessness. Let me ask you whether you favor an arranged marriage. No? Then perhaps I can tell you how it might otherwise happen. Quite by accident you meet a friend of a friend, or perhaps there is a brief exchange of glances and "suddenly you know that all through your lifetime...." For me one dance by chance at a freshman mixer made it all quite clear; a soft radiance of incomparable compassion and cute to the core.

Are those who feel that chance is so dreadful prepared to throw out all of the spontaneous moments, all of the unplanned events in their lives? When you visit Paris, be sure not to get up and casually swing open the French doors, step out onto the sunlit veranda, and savor the odor of fresh-baked baguettes below. And certainly don't look at the cart of dew-laden flowers briefly passing by to be stationed along the Seine River among the offhanded displays of watercolor paintings. And unless it has received a top recommendation in your guidebook, don't for a moment consider spontaneously approaching the saucy little bistro that caught your eye next door. You know the one; down there beside the boulangerie, with sweetheart roses on the red-checked table cloths and the waiter-cum-struggling-artist drying wine glasses beside the wheel of Camembert. It's just as chance arranged it—and it's all for you.

When I say that chance encounters provide some of the most romantic and satisfying experiences and opportunities in life, it does nothing to diminish the value of planning. We plan usefully. Lots of other animals plan. Of course, in the absence of a brain, flowering plants certainly don't plan. Or do they? It depends how you think about planning. Flowering plants have thrived in part because their perfumes have attracted pollinating insects like bees. The bees are rewarded by the nectar they drink and the pollen they collect on their legs. As the bees flit from flower to flower, they cross-pollinate the plants, leading to more vigorous seedlings. Mutual benefits are common in nature. It is wrong to see these as mere chance events. For much of nature the familiar phrase, "it was an accident waiting to happen" is quite appropriate in a positive sense. In glades filled with life, the odds of mutual benefit are reasonably high. Sure, the first encounter might be by chance, just as in romance, but there is selection for mutually positive interactions. Wherever organisms interact, it is easy to find examples of outcomes that are not so much at the whim of chance as they reflect an embrace of waiting benefits. There is an enormous amount of co-evolution of mutualism and cooperation in the life histories of plants and animals.

Life's intricate chemistry depends upon chance molecular encounters. Adequate molecular concentrations and strong affinities will assure statistically secure outcomes. As a practical example, when the adrenal gland atop the kidney releases adrenaline, it drifts around in the bloodstream until by chance some of the molecules encounter the heart muscle's specialized adrenaline receptors that strengthen the heartbeat during exercise.

Just as there was a significant element of chance in genetic changes throughout evolution's long history, so also there are significant elements of chance in the history of human affairs. Had John F. Kennedy been in the center of his PT boat when a Japanese destroyer rammed it, the United States would have had a different president. Had Marshall and Warren, Nobel Prize winners in Physiology or Medicine in 2005, not taken a few days of vacation, they might never have developed useful cultures of *Helicobacter pylori* that set them on their way to a true cure of duodenal and stomach ulcers. Is it credible that these chance-dependent historical outcomes were actually preordained; that these things were fated to happen, and were less desirable because they weren't?

Consider the role of chance in reproductive sociology and physiology. Do you feel your own personal existence was preordained? Who we are is a matter of chance. Naturally, we give thanks from time to time that we were lucky enough not to be born in some dusty village in abject poverty. And we laugh at the doctor's joke when we pass our health check-up, "Congratulations, you chose your parents wisely." Of course, we did no such thing. We are amazingly lucky to be born at all; we certainly had no say in it. Your appearance as an individual is astoundingly improbable. You would not be here without an unbroken lineage of chance matings over millions of years! If any one of the many thousands of links in our chain of ancestors had failed, we would not be here. That speaks eloquently to precision and reliability as much as to chance. To get a feel for these odds let's try to "design your own birth" starting with the long sociological odds. If my mother had not lost a roller skate she never would have gone skidding down the sidewalk in Austin to be swept into the rescuing arms of a handsome Texan walking nearby. All of us have similar romantic stories of unlikely meetings that led to matings. My wife's great-great-grandfather came from Germany to New York, and then continued by rowboat to a Michigan homestead. The plow horses were in top condition, but while he was busy furrowing his 160 acres, his neighbor was busy sowing with his wife. Being a practical German he made a deal with his amorous neighbor, "I'll trade my wife for your farm." Chance and necessity are never far apart. In time he had four sons by his new wife. The two eldest were inducted into the army to fight for the North in the American Civil War. They sent back formal yet sentimental letters now held by the Bentley Historical Library at the University of Michigan. Then the letters stopped coming. Both sons had been killed, two of the Civil War's 600,000 casualties. But the third son carried on a thin lineage by having one child who grew up to become my wife's father. If it is not enough that each of our births is up against such sociological odds, the odds against a specific birth become astronomically high when biology is also considered.

From among hundreds of your mother's eggs and millions of your father's sperm, only one chance pairing would ever give you. You are special; you are unique. When someone compliments you with the remark that you are one in a million, you should take that as a conservative estimate. You need only to go back a few generations to see how unlikely it is that you are here at all. The odds double for the previous generation because you needed a mating between your maternal grandparents to produce your mother and an independent mating between your paternal grandparents to produce your father. That is, a unique egg-sperm combination from each pair of grandparents was needed to have produced your mother and father. So the odds just going back to your grandparents are more than a trillion to one against your being born. You are the lucky winner of multiple sperm-egg lotteries. You should be enormously grateful for the opportunity to be alive, rather than egotistically asserting that all along destiny fated you to some pre-assigned role, whatever that might mean. Still, we can enjoy imagining that it was fate that brought two people together in romance. In tracing out the chain of human social interactions some people may be able to build a sense of destiny unfolding. But it is difficult to imagine what natural forces within a vast squadron of sperm could select a specific sperm as pre-destined to penetrate a specific egg of your mom in your turbulent beginnings. So let's be clear about your view. If you decline to see yourself as a product of the reproductive lottery, then do we conclude you see your particular existence as preordained? Do we explain the winning sperm in your case as divinely guided, somehow egged-on by the tiny hand of providence?

Evolution, God, Religion, and Heaven

It is one thing to label the Christian religion as sacred. It is quite another to label science as profane—as a profession devoid of spirituality. This is flat out wrong. There is ample reason for biological scientists to revere nature as they study the sacred core of life (Goodenough, 1998). Organisms and their DNA code are true living icons worthy of reverential feeling. Civilization is in the midst of just now gaining accurate information about the splendors of lives on Earth set in an accurate historical context. In contrast to the abundant pagan creation mythologies—errant relics of the past—the codes that give creatures their "endless forms most beautiful" are there for all to see. Research will reveal the majesty of the brain circuits that let us see color and form, feel a tender touch, and thrill to the rhythm and harmony of music. The brain provides us with prodigious feats of memory and reasoning, sublime feelings of ecstasy, and trust. Neuroscientists are coming to understand these as well.

Evolution provides a rational basis for a reverence for all life by ending the untenable claim that humans are special, vastly superior creations unrelated to other creatures. An understanding volution makes it arrogant to hold the self-serving view that God gave us dominion over all life because superior humans were set above baser creatures. To embrace the communality of life is to strengthen civil society. So if you cherish your pets and other animals, you should cherish the present revelations in molecular evolution that show our close genetic ties to other creatures. These are reasonable grounds for eliminating the unjustified suffering of permanently caged veal calves, of geese force-fed to bloat their livers for pate de foie gras, of de-beaked chickens, and of unanaesthetized bears repeatedly stabbed in the gall bladder for their bile. These creatures are too close to us to warrant such heartless abuse. It is our undeniable evolutionary relatedness that tells us so.

But you may still have a big question. "If I accept the evolution of life on Earth, does it mean science has ruled out a Designer, a.k.a. God?" To address this you need to recall the limitations of science and the scope of scientific research. Experimental science is a very practical endeavor. One can only measure what it is possible to measure. All scientists have many things they would like to measure or test, but can't figure out a way to do so. Every neuroscientist is interested in the relationship between the personal mental world and the physical world. Conscious awareness remains one of the great conundrums of science; it is not clear its biology will ever be understood. Scientists must be content with trying to explain nature in terms of physics, chemistry, and biology without invoking the supernatural. They cannot assign responsibility to supernatural forces which lie outside the reach of scientific methods. If a particular feature of biology is difficult to explain, biologists will wait for better methods or new facts that might contribute fresh understanding. The job of scientists is to frame questions in a way that allows testing. It is crucially important in science to design tests that will give compelling outcomes. Science has no direct way to examine the unexaminable, including a supernatural God. However, the more complex and capable we claim God to be, the more difficult it becomes to account for God's origins. God simply cannot be a vast uncaused complexity that instantaneously arose from nothing. Yet, it is anathema to traditional theology to suggest that the principles of evolution are a reasonable way to account for the presence of a highly complex God, or even that before Creation, God had to mature from a once helpless, motherless babe to a sophisticated God the Father, much as Jesus went from babe to man on Earth.

Fearing Moral Decay and Cultural Chaos

Is it dangerous to believe in evolution? Will widespread acceptance of evolution somehow precipitate a terrible social descent into a coarse and immoral society? This is a concern worth addressing more fully. We can begin with *Mere Christianity*, that slender essay by C. S. Lewis that has persuaded so many people of Christ's way. In his own conversion from atheism to Christianity, Lewis argued that a loving God must exist because humans have an innate sense of what is moral, a natural feeling for right and wrong. However, if humans are born with a uniform moral compass, an absolute sense of moral correctness, why don't all Christians adhere to closely similar standards? One needs to explain, for instance, why Christian thought includes both those who totally reject the death penalty as murderous and others who find execution a fitting fate for homosexuals or even for those who work on the Sabbath. For some the moral compass points north, for others south. In any event it is unlikely that mere belief in biological evolution will shred the moral raiment that cloaks culture. The moral fabric of society is not so easily frayed; it is a robust weaving of many threads of nature and nurture. Evolution is compatible with a firm moral sense.

Re-examinations of the possible origins of human cooperation and altruism suggest biological roots (Bowles, 2006; Nowak, 2006). In recognizing a moral rule of reciprocity, epitomized by the Golden Rule, we are bound together by an intricate legal system that offers the hope of justice. We have elegant documents and declarations of human rights, ideals of universal education and suffrage, freedom of speech and freedom from coercion, networks of obligations to charities and churches, friends and family—these are just some of the threads woven into the fabric of civil society. If only leaders enthralled with power and haughty religiosity would honor these values.

Is naturalism harmful? Is it really likely that broad acceptance of naturalism (we can take this to mean an absence of supernatural influences) will so erode the base of morality that "society would come apart in a flash", as Bruce Chapman (2000), the president of the Discovery Institute says? If so, how is it that the majority of Europeans have already embraced evolution without tumbling into a moral abyss? Neither in the United States nor in Europe does the evidence support the speculation that the acceptance of evolution means a decline in morality (Paul, 2005). The Japanese accept evolution while having the lowest rates of homicide, suicide, and child mortality, and the greatest longevity, yet they are the least religious—only 10% believe in God. The United States stands apart from other developed nations because with the highest religiosity and the lowest belief in evolution it has the highest rates of homicide, infant mortality (along with Portugal), and the highest incidence of adolescent gonorrhea and syphilis, pre-marital pregnancy (15-17 years), and abortions (15-19 years). Contrary to those who lambaste evolution, these survey results suggest that a high incidence of religiosity is neither necessary (Japan) nor sufficient (United States) for societal health. The moral behavior of America's youth is likely to be improved by a better grasp of biology and evolution, not less.

A wider comparison across countries helps to reveal the relationships between religiosity and social variables. Paul (2005) has made a preliminary assessment of data from a survey of Portugal and 17 rather more developed nations that included Australia and New Zealand, Canada, United States, Japan, and most European nations (International Social Survey Program). Religiosity indicated by a belief in God, church attendance, praying, and biblical literalism did not decrease the rates of homicides, youth suicides (15-24 years), the incidence of adolescent gonorrhea and syphilis (15-19 years). Nor did religiosity alter life expectancy. However, the rates of infant mortality (0-5 years) and abortion among adolescents were higher among the religious. In contrast, infant survival was positively correlated with acceptance of the theory of biological evolution. At the least, these correlations show that religiosity is no runaway winner in the quest for higher morality.

Material decadence? In many nations the human condition has markedly improved in the last century. The rising standard of living is associated with a reduction in religious practice. In country after country in Western Europe religious participation decreased as the standard of living increased from 1970 to 1999 (Norris and Inglehart, 2004). In contrast, populations that remain vulnerable to risk and uncertainty "regard religion as far more important." As education, per capita income, and increases in longevity combine to lift the quality of life, interest in religion fades. Reduced religiosity seems more closely tied to a better standard of living than to a greater acceptance of evolution. Chapman (2000) has claimed that, "materialism is a major source of the demoralization of the twentieth century...[and] has done untold damage to the normative legacy of Judeo-Christian ethics." Are we to conclude the better off are morally depraved?

In combination with a higher standard of living there is greater attention to nutrition, child welfare, and human rights, with social security available for the elderly, and medical services for the infirm. These efforts represent improved sensitivity to the human condition. This is clear moral progress. On the other hand in the United States sexual mores are lax, child pornography prevalent, and divorce rates are high even among born-again Christians. The data do not support Chapman's efforts to pin immoral behavior on naturalism and materialism.

Some critics blame science and technology for an assortment of ills: for coarsening culture, for creating or exacerbating the problems of abortion, assisted suicide, contraception and the morning-after pill, designer drugs, pornography, sexual permissiveness and crime, and the violent mayhem of modern warfare, including the triple threat of biological, chemical, and nuclear war. In explaining degenerate public morals one might better scrutinize the commercial advertising and entertainment industries, where the media—its producers and glittering stars—are vastly more responsible than science for debasing sexual conduct and promoting drugs and violence. With more free time and opportunity to engage in the classic deadly sins we are vulnerable to commercial manipulations (Huesmann, 2007).

Insecure fundamentalists may trump up charges of social decay because they personally feel their Salvation and special standing are threatened by what we know about the evolution of life on Earth. In any case, it is still proper to ask whether those of faith or those who claim to be doing God's will (apart from religious employees) are more ethical and less selfish that those who believe in evolution. Every institutional religion is proud to show off the moral superiority of its members where possible. However, there is no evidence that the non-religious commit more crimes. Thus, non-religious individuals are not imprisoned in greater proportions than the religious. It would require another book to explore in detail whether civil behavior, better health, improved stress reduction, a cheery outlook, and additional positive personal traits are consistent benefits of religious belief.

Any scientist who contributes even indirectly to the development of destructive weapons must accept some responsibility for their injurious use by others. Of course, those who exploit scientific understanding to build and unleash destructive weapons against others bear enormously more responsibility. For example, a visitor to our planet would have some difficulties telling who the bad guys are in ongoing conflicts. Aren't the bad guys the ones who torture suspects, slay children, and mutilate civilians with weapons like bombs and land mines? When flesh and blood are splattered, for a good cause or bad, you need to look at those who hold power and the wealth used to influence it—they will not be the scientists.

We are now poised to learn which particular genes or regulatory sequences of DNA distinguish humans from chimpanzees, our closest living relatives, and from Neanderthals, our much closer, recently extinct relatives. It will also become apparent whether these differences are a reasonable outcome of evolutionary processes or are so unique and distinctive as to suggest an imposed Intelligent Design.

15

&

Intelligent Design: Adrift Toward Gods?

Seeking Evidence for God in the Natural World

Creationists and scientists offer notably different narratives to explain the same wonders of nature. Biologists argue persuasively that the intricate complexities of biological adaptations do not require an Intelligent Designer (Dawkins, 1986). As its alternative to the evolutionary explanations provided by geology and biology, the 20th century movement of "Creation Science" embraced the biblical stories of Genesis. In order to gain access to public classrooms, creationists had hoped to attach the Old Testament Creation story in Genesis to the teaching of science. Preferably creationism would replace the entire scientific account of the evolution of life on Earth. However, in Edwards v. Aguillard, 1987, the United States Supreme Court held that the intrusion of creationism into public schoolrooms was an unconstitutional breech of the separation of church and state. Creation Science was declared to be religious dogma rather than science. While this ruling temporarily freed science teachers from religious intrusions, it re-invigorated the determined supporters of divine Creation. Once again they saddled up, hoping to breech schoolroom walls of separation with cleverly deployed Trojan horses.

In the Marketing of Supernatural Explanations, "Creation Science" Has Been Reincarnated as "Intelligent Design."

The successor to the creationist religious doctrine is Intelligent Design (ID), which tries to mask its theology by deceptively representing itself as a scientific exercise. Fundamentalist proponents of Intelligent Design claim it is science, not on the basis of their experiments and data in properly peer-reviewed research publications, of which there are none, but rather on the grounds that ID offers the only reasonable interpretation for a few biological systems hand-picked by ID for their impact in public relations campaigns. The proponents of ID claim that half a dozen biological systems like a living cell, or blood clotting, or the spinning flagellum that propels a swimming bacterium are each too complex to have evolved naturally. They see Intelligent Design as proof of God; just as more than seven centuries ago the pre-eminent Catholic theologian Thomas Aquinas (1225-1274) saw design in nature as evidence for the existence of God.

Common Sense is a Flawed Guide for the Origins of Biological Complexity

We can begin with the analysis of the Reverend William Paley (1743-1805), who put the argument well in 1802. While he was out for a stroll, Paley imagined that if he had discovered a pocket watch on the ground, its complexity would strongly suggest that the watch had a designer. Paley and his acolytes went on to claim the human eye must also have had a Designer, because in contemporary phrasing it was irreducibly complex. They alleged that an organ as complex as the human eye could not have evolved as an accumulation of numerous small steps spread over many thousands of years. It was argued that intermediate visual detectors, like a patch of light-sensitive skin, would not have been retained over successive generations because they would have been useless. Instead, God must have designed and built the human eye all at once, along with creating all of the other creatures on Earth.

Meanwhile on the other side of the Atlantic Ocean also in 1802, Thomas Jefferson was expressing somewhat different thoughts in a "Letter to the Danbury Baptists." Thomas Jefferson stated that his aim behind the sentence in the first amendment to the United States Constitution, "Congress shall make no law respecting an establishment of religion, or prohibiting the free exercise thereof," was to "build a wall of separation between church and state." There are constant challenges to the protections afforded by this wall.

Beginning forcefully in the 1990s a small group of American Christians reactivated the Intelligent Design argument. However, they steadfastly refuse to credit Reverend Paley's lead. They want to hide their religious beliefs and objectives in order to masquerade as a scientific alternative to evolution. ID's strategy is to convince politicians, school boards, the lay public, and the courts that Intelligent Design is a better "scientific" explanation than evolution for some of the remaining gaps in our understanding. They ask school boards to insist that "scientific" criticisms of evolution be given a fair hearing. "Teach the controversy," they argue. They appeal to Americans' sense of fairness by casting themselves as a beleaguered underdog in a society dominated by science, even though the majority of Americans do not believe in evolution. Step-by-step, state-by-state, it is ID's complex design to have sympathetic local school boards open the doors of American science classrooms to ID's Trojan horse. In this, their "Entering Wedge Strategy" (Johnson, 2002), the Trojan horse contains no experiments, or scientific data, or scientific research publications, just their Intelligent Designer, the Christian God, along with a hidden stash of Bibles, crosses, and other religious icons for later distribution in classrooms. Has ID really found that biological systems have scientifically inexplicable features or is ID flawed? Here are eleven flaws in the Intelligent Design proposal.

A First Flaw in the Formulation of Intelligent Design: ID Depends Upon the Argument of Common Sense

Proponents of ID appeal to common sense in speculating that complex biological systems could not arise by slow evolutionary changes. However, professional biologists who study evolution have come to a different conclusion. Common sense is defined as an on-the-spot judgment that relies on the elements of understanding gained through experience. Although common sense can be a helpful guide for suitably experienced individuals, it often fails in technical matters. Throw a heavy rock into a pond and, to no one's surprise, it will sink like a stone. Relying on such experiences, one's common sense would also suggest that a heavy iron boat would quickly sink. But an intact concrete or iron boat will float if it weighs less than the water it displaces. Archimedes' principle of buoyancy allows all of today's large ships to be welded together from enormous steel plates that if individually plopped into the water would immediately plunge to the bottom of the sea.

Common sense would suggest that after American astronauts inadvertently left some *Streptococcus* bacteria on the moon, the marooned bacteria would surely have died after two years without air or water. But they survived. Indeed, really cold bacteria presently frozen in Siberian tundra may have survived for three million years. Even if rare thaws of the tundra occasionally revived the bacteria and allowed some renewal through replication, there is still an impressive period of survival under freezing conditions. A cold bacterium might divide only once a century, if that. Under warmer, more favorable conditions, rapidly dividing bacteria could generate about 300 trillion descendants in one day, provided they had adequate food and space. Some bacteria are so tough they can endure the pressure of the ocean depths, inhabit crevices in rocks thousands of feet deep in the earth, and survive in other harsh environments including boiling mud and alkaline salt flats. Bacteria can eat meals of manganese and iron, or wood, paint, crude oil, and even sulfuric acid. Talk about cast-iron "stomachs." So common sense isn't much of a guide to the capabilities of bacteria or to early evolution when ancestors of bacteria were the first cells to evolve on Earth before there were any multicellular plants or animals.

Common sense also tells us that we have the same skin, muscles, and other organs we had 5 or 10 years ago. Yet, over a short period most of your tissues and other molecular parts will be renewed. Not only are the layers of epithelial cells in your skin and the lining of the intestine replaced during the course of a week or so, but most of the other organic molecules throughout your body will be replaced in the course of months or a few years at the most. Bio-molecules are unstable. Damaged molecules are replaced. Other molecules or entire cells are on a fixed schedule of replacement—like a scheduled oil change for a car. The periodic replacement of molecules and cells keeps our tissues in working order.

These counter-intuitive examples of concrete or iron boats that don't sink, of the prolonged survival of rugged bacteria, and our own molecular renewal indicate the importance of actually doing science—of the need for reliable observations and experiments to avoid errors from guessing. The conclusion is clear: basing important judgments on common sense can lead you astray. Deficient in data, ID takes the unpersuasive approach of relying on appeals to common sense and plausibility including its core notion that complexity requires a designer.

A Second Flaw in ID: Does Irreducible Complexity Exclude Stepwise Origins?

As they examine diverse biological systems, the advocates of ID hope to ferret out a few systems with two coveted properties: the system must not yet be scientifically well understood and it must appear at first glance to be irreducibly complex. Does the irreducible complexity of the human heart mean it must have been created all at once? ID proponents speculate that hearts could not have arisen in a step-by-step fashion because the intermediate stages would have been functionless. Let's look at the scientific findings.

In order to function as an effective pump the mammalian heart clearly needs valves and chambers as well as leak-free connections to various arteries and veins. However, the evidence is good that the four-chambered heart of birds and mammals began as a two-chambered heart, and evolved in stages to the three-chambered heart of amphibians, to reptilian hearts with four chambers, ultimately further specializing in birds and in mammals. Indeed some of this evolutionary progression in the structure of vertebrate hearts is mimicked by the mammalian heart's progressive embryological development. Each developmental step, from two to three to four heart chambers, is both functional and irreducibly complex. After all, the failure of any component of a two-stage pump is potentially fatal for the embryo.

The flagella of eubacteria are supposed to be irreducibly complex organelles whose proteins were miraculously created all at once just for propelling the bacterium. It is puzzling that this example continues to be used by ID advocates when it is known that at least ten of the flagellar proteins have long histories of other functions in bacteria. Perhaps the advocates of Intelligent Design are anxious to keep it a secret that the black plague bacterium, *Yersinia pestis*, uses some of these proteins as deadly toxins. Natural selection is opportunistic in building structures from already existing components.

A Third Flaw in ID: Was the First Cell Built All at Once?

Let us consider the claim that cells are irreducibly complex. No doubt about it, cells are complicated. Many of their parts, their organelles, depend upon each other. Destroy any major organelle and the cell is likely to die. Does this apparent irreducible complexity also imply that a cell must have been created all at once? Not at all. It is just as easy to imagine that cells evolved by adding features here and there, which were later yoked together. Our present limited understanding about the evolution of cells is precisely why some ID advocates spring to advertise the cell as evidence for an Intelligent Designer who personally built the cell from scratch—a "God of Gaps" in our scientific understanding. Yet, biologists know enough to identify some remarkable stages in the build-up of cells. There is abundant evidence that several organelles present in today's cells are the remnants of bacteria-like microbes that eons ago managed to invade cells and set up residence. The most celebrated examples are the chloroplasts in photosynthetic plants and the mitochondria that are the power stations of all advanced cells. Mitochondria are derived from tiny unicellular organisms that originated more than a billion years ago. Perhaps beginning as parasitic invaders, mitochondria survive today as pared down remnants—sources of the energy our cells now need to survive. Mitochondrial enzymes oxidize the sugar glucose to make ATP, the molecule that stores most of the cell's instant energy. We couldn't live without the numerous mitochondria that each of us inherits solely from our mothers. In addition, multiple bacterial genes have combined with cellular DNA to extend the repertoire of biochemical reactions. Thus, although the detailed steps of cellular evolution remain unknown, the infusion of microbial DNAs is part of the evidence that today's complex cells arose in stages rather than all at once. Advantages that were initially discretionary eventually became more essential or "irreducible".

A Fourth Flaw in ID: Can Gene Duplication Give Rise To New Roles?

Within the span of 100 million years every gene of a long lasting multicellular species can be expected to duplicate. With two copies a gene can generate roughly twice as much of a useful protein. But a more interesting outcome is the prolonged transformation of the second copy into a new and ultimately essential role. As part of the evolution of the mammalian eye, gene duplication helps to explain our vision for different colors (see the eighth flaw).

A Fifth Flaw in ID: What is the Probability That Occasional Changes Could Generate a Complex System?

A popular effort to dismiss evolutionary explanations is to cite terribly long odds of assembling complex biological systems by chance alone. Most assuredly it is highly unlikely that a complex biological system would arise if several genetic changes had to occur simultaneously within one individual. Since that is unlikely, ID then claims the only other option is to have a mysterious, superior intelligence suddenly make the complex system as a whole with all of its features miraculously created simultaneously. However, in many instances one feature is already in place, so the odds of having two are merely the odds that this second feature will arise. If this added feature improves an organism's chances of survival and reproduction, it is likely to be retained as part of the mix. Many generations later the two might interact in a different, perhaps more complex, application. The important point is that along the path leading to a more complex design, intermediate features were functional, although often in ways unrelated to their latest role. This means that novel complex systems can arise by linking existing useful modules. For example, before blood clotting evolved, several of these proteins were already serving useful functions wholly unrelated to blood clotting (Doolittle, 1993). Some of the multiple blood factor proteins are related to one another because they are derived from one primordial gene. The quirks of chromosomal rearrangements shuffled together pieces of primordial genes in different combinations, including the eight dispersed sequences of bases that now make up the coding portion of the gene for a blood clotting protein, factor IX (Stamatoyannopoulous et al, 2001).

Moreover, ID proponents overestimate the number of changes required to create a novel complex system. For example, it may take only a few changes in regulatory sequences of core processes to generate major changes in the form and function of forelimbs. When biologists study complex systems, they encounter familiar core modules. Recent research in molecular biology suggests that contemporary organisms share perhaps a dozen core biochemical processes (Kirschner and Gerhart, 2005). Altered regulation of various combinations of these core processes gives rise to new organisms. A module used earlier in one context may be re-deployed in a new context where it performs a useful function. That improves survival.

A Sixth Flaw in ID: Fossil Gaps Do Not Undercut the Continuity of Evolution.

You have probably seen the series of cartoon pictures of vertebrate evolution in which a fish-like creature slithers out of the sea and begins its walk up the beach as a stubby-legged amphibian. As it reaches the dry sand dunes, it looks more like a reptile, and then further onward it resembles a monkey, then a knuckle-walking ape, and finally a primitive slouching human. Such imaginative depictions artfully compact millions of years of "ascent" with modification into a stroll up to the high, dry beach. Many critics have claimed that gaps in the fossil record undercut any such progression, scoffing in particular at the demanding change from a fish with gills and fins to an amphibian with lungs and legs. This criticism was deflated by a remarkable fossil discovery by Coates and Clark (1991). They found the fossil remains of an amphibian whose gills allowed life in the sea, while its four legs and lungs allowed it to walk and breathe on land—a long-sought transitional form. But a fish should not shift into an amphibian all at once. So where is the evidence for more transitional steps between fish and land vertebrates?

More out of drama than logic we like to imagine "missing links"—those oh-so-desired fossils that provide evidence for the transition to a major change in habitat or function. The evolutionary reality is not quite so emotionally singular because major transitions probably reflect a set of small changes spanning a series of species. The proto-bird *Archaeopteryx* is an icon of evolutionary biology in the transition between reptiles and birds, but there were surely additional intermediates.

If you want to explore the habitat of ancient vertebrates associated with the transition from water to land, I recommend that you pack some mosquito repellent and trek to the ancient river flats on Ellesmere Island in the far north of the Canadian Arctic. It was there that Neil Shubin of the University of Chicago and his colleagues sought transitional fossils for three fruitless summers. Persevering for a fourth year, they found a fossil mother lode of *Tiktaalik roseae* that included two intact skulls attached to part of the body (Shubin et al., 2006). These recently discovered fossils of this crocodile-like animal may become another evolutionary icon, one of the long-sought intermediates between fish and land vertebrates or tetrapods. When the Canadian far north enjoyed tropical conditions during the Devonian period some 380 million years ago, mutations led to *Tiktaalik* equipped with stubby forelegs able to walk in the shallows. John Maisey of the American Museum of Natural History in New York City has described *Tiktaalik* as "the most significant discovery in years." This transitional form makes it evident that land vertebrates evolved from aquatic vertebrates (Fig. 15.1).

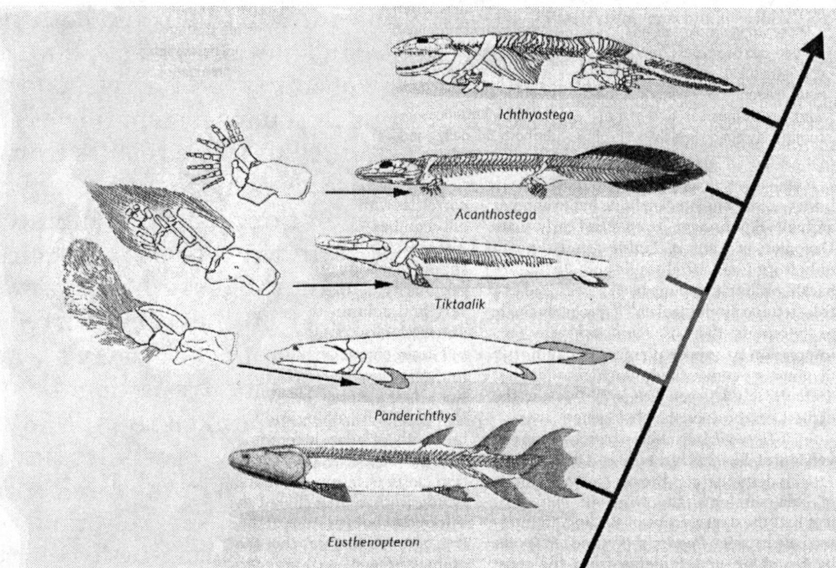

Figure 15.1 Until recently the known fossil record has had a substantial gap between fish-like fossils (example: *Eusthenopteron*) and land-based four-legged vertebrates. The recently discovered crocodile-like fossil, *Tiktaalik roseae*, is a new species that helps to bridge the transition from a primitive intermediate (*Panderichthys*, 385 million years ago) to land-like fossils (*Acanthostega* and *Icthyostega*, 365 million years ago). The transition toward walking legs is most evident in the bones and fine rays of the pectoral fins shown at the left for *Panderichthys*, *Tiktaalik*, and *Acanthostega*. *Tiktaalik* has a mixture of finger-like bones and terminal fin rays. The landform *Acanthostega* has no delicate fin rays. (After Ahlberg and Clack, 2006 and Shubin et al., 2006.)

Well, what about whales? Where are the fossils supporting the claim of a reverse transition from land to water? Even though whales, and their 80 or so kin, frolic in the ocean depths, they are air-breathing mammals that give birth to live young, which they nurse with their milk. Comparisons of DNA suggested that whales were related to large land mammals like the hippopotamus (Nikaido et al., 1999). Even so, creationists insisted on transitional fossils as proof that land-dwelling mammalian ancestors of whales actually walked from the land to take up residence in the sea. Even the great evolutionary biologist, George Gaylord Simpson, called whales the most "aberrant of mammals."

The recent discoveries of transitional fossils of terrestrial whales are a landmark proof of this land-to-sea evolutionary transition. *Ambulocetus natans* was a whale that could both swim and walk—its leg bones tell the tale (Thewissen et al., 2001). As further anatomical evidence, the special type of anklebone that helps hippos walk was also present in the fossil remains of early whales that had legs (Gingerich et al., 2001). The results from the distribution of the different oxygen isotopes in fresh and seawater are also persuasive. The lighter form of oxygen predominates in fresh water ($^{16}0$) while the heavier form of oxygen is more prevalent in seawater ($^{18}0$). And sure enough, the fossil bones of whale ancestors have higher concentrations of $^{16}0$ like other terrestrial animals in contrast to the predominance of $^{18}0$ in today's whales living in seawater (Thewissen). In finding transitional fossils with legs, scientific investigations have debunked the speculation of creationists that the mammalian ancestors of whales have always lived in the sea. By filling in major gaps in the fossil record, the recent discovery of more "missing links" severely undercuts the traditional argument that the absence of bridging fossils rules against the smooth and continuous evolution of species.

A Seventh Flaw in ID: Why Irreducible Complexity is Unintelligent Designing.

The uncanny ability of members of ID's inner circle to recognize the brilliance of the Designer's handiwork in the irreducible complexity of biological designs leads to some self-contradictory claims. Engineers intentionally build in redundancy to prevent the breakage of a single component from shutting down the entire system. An intelligently engineered system will have a fail-safe capability that allows it to continue to operate after one component malfunctions. It would be an unintelligent engineering design to make a system irreducibly complex, one that would quit working with the failure of a significant component. To claim biological designs are irreducibly complex would seem to insult the intelligence of the Designer, whoever that might be.

An Eighth Flaw in ID: A Long Parade of Unintelligent Designs

Evolution produces many designs. Some complex biological systems are so lacking in elegance or utility, they would embarrass any self-respecting designer.

Some functional and some dysfunctional outcomes of gene duplication. There are three types of color receptors in the human eye having maximal sensitivity to red, green, or blue light, respectively. The gene responsible for the blue absorbing pigment arose first. Duplications of the "blue" gene and later modifications produced the "red" gene and the "green" gene that lie next to each other on the X chromosome. The "red" and the "green" proteins have remained quite similar, differing in only 15 out of their 348 amino acids. During chromosome duplication in the ovaries the two X chromosomes may not separate properly. As a result the one X chromosome normally present in male embryonic cells may lack the green gene or alternatively may gain some or all of a second green gene. The X chromosome of more than half of all men has at least two genes for the green sensitive pigment. About 1% of men lack the red gene and 2% lack the green gene. For such individuals red and green may both look like a muddy yellow. Is God to be saddled with responsibility for these errors in complex designs that cause anomalous color vision?

Pseudogenes are tangible records of abandoned solutions. Our DNA is sprinkled with hundreds of degraded genes called pseudogenes. Many pseudogenes reflect gene duplication events. Pseudogenes are genes that were probably once functional, but now are useless relics that remain in the strands of our DNA. These dregs are discarded steps in the evolution of biochemical pathways. The old functions they carried out are handled by newer mechanisms, often by more intricately regulated biochemical pathways. Pseudogenes are obsolescent residues replaced by the stepwise build-up of better systems. Why wouldn't an Intelligent Designer have made a clean and elegant design at the outset, free of pseudogenes that now litter our chromosomal DNA?

A Ninth Flaw in ID: Does the Designer Have Flaws?

Extinct designs are failed designs. Let's assume life on Earth is the handiwork of an Intelligent Designer. Advocates on behalf of the Designer point to certain species and certain biological structures as designed. Since an entire book put *Darwin on Trial* (Johnson, 1993), we can follow the rules of fair jurisprudence and request an opportunity for rebuttal. Let's question Johnson as the designated speaker for the design engineer. "Why did so many designs fail?" For example, 20 of 22 species of elephants have become extinct in the last few million years. Why? The extinction of the woolly mammoth was so recent that entire specimens have been found preserved in icy bogs. Now gone, these species will never reoccur in the wild. If survival is the hallmark of a good design, such track-records reveal a predominance of flawed designs that resulted in the extinction of the vast majority of species—there are few survivors from ancient times. Does this mean there were problems in the design or in the workmanship? Hopefully ID proponents have some explanation beyond trial-and-error designs or a failure to adjust to environmental changes, for those smack of evolution.

The Designer was inordinately delayed in delivering multicellular designs. If humans were destined to have the most special place among all of the glorious forms of life the Designer created, why were there only single-cell organisms for the better part of two billion years? Why did it take another billion years of designing multicellular organisms before any large animals were made? Is it because it was so difficult a task that it required many trials and errors? Shouldn't we have expected perfection from the outset? Why were humans created in the last eye-blink of time? Is this because the millions of species created in the first three billion years were imperfect proto-designs allowed to die off? Was extensive practice needed to make humans? Even now few humans feel like they are the very peak of design perfection. Was it by intent or by error that our backs ache, our eyes need corrective lenses, our noses and lungs suffer hay fever and asthma, and we fall prey to depression or various addictions? Some of us would have welcomed more adept, robust designs.

A Tenth Flaw in ID: Are There Multiple Intelligent Designers with Differing Skills?

In the real world of engineering, multiple designers construct complex systems. It is a humbling experience to examine an elegant and complex mechanical device like a watch with its jeweled moving wheels and tiny stepper pins, or to hold a circuit board with its intricate and neatly crafted designs—the electronically pulsing heart of modern computers. Did one individual invent this level of intricacy and detail? In the real world computer circuits and watches were neither developed by one person nor made from scratch. Inventions progress though a series of prototypes, as engineers test suggestion after suggestion for possible improvement. Examinations of the history of engineering design and product development reveal a formal resemblance to the likely pattern of the evolutionary development of living designs. The elegant machines of today have emerged from dozens of failures along the way. The beautifully crafted mechanisms of a finished design reflect hard-won, step-by-step improvements over the course of the machine's evolution. Engineers will try a change here, a mutation there. If, for whatever reason some change makes a system worse, that model is discarded. Consider the history of artificial heart design—dozens of modifications on model after improved model, yet difficulties remain. In the fullness of time bioengineers will build a reliable and trustworthy heart pump that overcomes nagging problems like mechanically crushed blood cells and the formation of blood clots that can lead to lung embolisms or strokes. Success will eventually come following a long series of small incremental changes in design, a few of which survive as improvements while most are discarded after their testing reveals problems.

So, if a complex biological system, like a cell or our eyes, were actually a designed system, the designer is much more likely to have been a team of designers. The best strategy for constructing an advanced eye would be to have one design group work on color detection, another on motion detection, still another on the detection of geometric forms, and so forth. Of course, if one is precommitted to the less probable alternative of a single designer, it is possible to dream of a supernatural intelligence, so magical and magisterial that it built the entire human eye from scratch. ID cannot abide by design through the much more credible alternative of multiple designers, because multiple Gods would rule out the vaunted single god they never mention but fervently wish to introduce into science classrooms. There is but one non-negotiable script for the Trojan horse of Intelligent Design. It will be rolled into science classrooms across America to mark the triumphal entry of a solitary male Christian God.

An Eleventh Flaw in Intelligent Design: The Malevolent Designer—A Maker of Bad Deals

Why do Intelligent Design advocates only consider beneficial biological systems? Why do they ignore nature's many complex designs that are malevolent? As Elaine Pagels (1995) argued in *The Origin of Satan*, satanic influences began millennia ago as mean spirits and ill winds—frightful cold drafts of sickness and harm. In time, satanic spirits coalesced into an evil person who could be more easily imagined and derided. In order to explain the persistence and considerable complexity of certain cruel acts, it became helpful to elevate Satan to a scheming and cunning role. Does the Discovery Institute, the political base of ID, hold Satan responsible for Malevolent Design? Can they escape holding God accountable for irreducibly complex malevolent designs by foisting responsibility onto an Intelligent Maligner? For example, immune cells, able to recognize and attack harmful invading microbes, guard the gateways to our bodies. Nonetheless, the fiendish AIDS virus repeatedly transforms itself into varieties that elude the sentries protecting the portals of our lives. In shielding God, the members of Discovery Institute could attribute AIDS and many other examples of Malevolent Design to the complex trickery of Satan who on numerous occasions has transformed himself into ordinary creatures or even held innocent victims in bondage to his purpose-driven strife. Classes in religion could "teach the controversy" between Satans and Gods as strife evidently continues in the form of Intelligent Maligners pitted against the Intelligent Designers.

The Aims and Operations of the Intelligent Design Campaign

The "Intelligent Design" movement is primarily a well-funded public relations campaign orchestrated by the Discovery Institute in Seattle. The Discovery Institute spends its considerable financial resources on political lobbying, buttressed by loosely argued anti-scientific books and pamphlets. It aims to replace the explanatory power of biological evolution with conservative religious views in science classrooms. Gussied up for prime time, ID is born-again Creationist Religion that has no legitimate place in science classrooms. Indeed, ID doesn't even warrant a place in classes on comparative religion, for where is its body of religious beliefs and doctrine that can be openly discussed? As its leaders have repeatedly made clear, ID is a rapier intended to skewer evolution in order to slip a solitary God past the United States Constitution that guards American public schoolrooms. ID gains traction by arousing public longing for evidence that God exists, by creating and exploiting misunderstandings of evolution to evoke visceral fears, and by specious pleas that science classrooms should be opened up to trumped up controversies which are entirely religious formulations.

Public Science Education and the Law

In the 1930s and 1940s the coverage of evolution in high school textbooks was progressively reduced as publishers found that book sales rose when they slighted the scientific information about evolution. Yet, by the 1950s evolution had unquestionably become the most important organizing principle for understanding life on Earth. With the backing of various individuals and organizations, evolution again became more prominent in textbooks. Ironically, students have seemed indifferent to these swings of emphasis. Even today, half of American adults still do not accept biological evolution (Harris Polls, 1999 and 2005). Such ignorance in a technologically advanced nation is dismaying for it undercuts a democratic society that depends upon an informed public to support rational approaches to biomedical issues. As a consequence, some uninformed school boards have behaved with "breathtaking inanity," according to a recent court decision in Dover, PA.

On December 20, 2005 Judge John E. Jones III released his decision in Kitzmiller et al. v. Dover Area school board. This first legal test of ID produced a meticulously argued, stinging denunciation of attempts to introduce religion into the science classrooms of public schools. Here we consider some highlights of this 139-page decision.

The courts have interpreted the Establishment Clause of the First Amendment to the United States Constitution to mean that regardless of its stated intentions, a government agency must not convey a message of either endorsement or disapproval of religion. According to creationists "...all scientific evidence which fails to support the theory of evolution is necessarily scientific evidence in support of creationism ..." (McLean decision of the US District Court in Arkansas, 1982). Yet, the McLean decision held that in limiting the contending explanations of life's diversity to either biblical creationism [renamed Intelligent Design] or to evolution was to set up a "contrived dualism." By allowing just these two choices, creationists could then claim that "one must either accept the literal interpretation of Genesis or else believe in the godless system of evolution—those are your only options." In Epperson, 1987 the Supreme Court agreed this was a false dichotomy and further held it was a constitutional violation to forbid the teaching of evolution for religious reasons.

Recently, the community of York (Dover), Pennsylvania was energized by its local School Board's promotion of religion in public schools. The town newspapers, the York Daily Record and the York Dispatch, published 225 letters to the Editor and 62 editorials on the subject. The 2005 Kitzmiller decision handed down by Judge Jones stated, The Dover School Board's disclaimer statement to be read in classrooms "...singles out the theory of evolution for special treatment, misrepresents its validity without scientific justification, presents students with a religious alternative masquerading as a scientific theory, directs them to consult a creationist text as though it were a science resource, and instructs students to forego scientific inquiry in the public school classroom and instead to seek out religious instruction elsewhere." It is a dangerous and specious policy to force students to choose between God and science. Judge Jones further wrote, "Both Defendants and many of the leading proponents of ID make a bed rock assumption which is utterly false. Their presupposition is that evolutionary theory is antithetical to a belief in the existence of a supreme being and to religion in general."

Mainstream Christian denominations in the United States actively opposed to Intelligent Design include: the Episcopal Church, the Lutheran World Federation, the Presbyterian Church, the United Church of Christ, and the United Methodist Church.

16

&

Fundamentalism and Faith

Science is not a Faith-Based Endeavor

No one has characterized the two elements of faith with more grace and elegance than Paul. "Faith is the substance of things hoped for, the evidence of things not seen" (*Paul, Letter to the Hebrews* 11:1*)*. In this chapter I focus on the problems generated by faith, rather than the vaunted emotional reassurances provided by faith.

Science deals with measurable features of the workings of natural events. Particular scientific truths or relationships are established by observations, experiments, calculations, and logical inferences, not by a faith in the existence of what our hearts desire. It is sometimes argued that by assuming a real and orderly universe, science reveals its own form of faith. It is more accurate to say that a scientist's belief that events have discrete causes reveals confidence that additional lawful relationships can be discovered, given the millions of regularities already demonstrated. This is "faith" in the sense of a "general confidence" based upon proven principles, not "Faith" in the religious sense of attempting to transmute one's desires into factual truths. Scientists have a general expectation that further experiments will broaden understanding. At the end of the day it is the securing of hard-won findings rather than a commitment to faith that reveals how lawful principles govern nature.

Religious doctrines often deal with unsolvable mysteries that characteristically involve a supernatural world. Many individuals have undergone a personal religious transformation that gives them the sense of experiencing a supernatural God. They claim that feelings this strong and pure can't be wrong. Yet, are intense feelings sufficient to establish or reveal truth? It would be more persuasive if those wedded to the accuracy of their faith could offer up solid predictions of future events as proof they are on to something. Regrettably, the intensification of faith yields no increase in accuracy.

Secular Humanism is not Religion

"Secular" means worldly or lacking in supernatural beliefs. Nonetheless, after a U.S. Supreme Court opinion lumped secular humanism together with Buddhism and Taoism as godless systems (Torcaso v. Watkins, 1961), some individuals wrongly tried to label humanism as yet another religion that should be quarantined from influencing the United States government. However, the defining requirement of religion and religious belief is the inclusion of the supernatural realm (God, angels, saints, divine answers to prayers, Heaven, reincarnated souls, and so forth), whereas humanism, like science, deals only with the natural or material world. Since secular humanism makes no supernatural claims, it fails to meet this essential requirement of a religion. Humanism implies concern for human welfare, peace, prosperity, and liberty in an open, non-totalitarian society. Secular humanists are uncomfortable with setting forth more detailed doctrinal beliefs which might limit personal viewpoints.

Western Religious Fundamentalists

Those who believe that the basic religious truths are unchanging may embrace religious fundamentalism, including its opposition to many beliefs of secular "western culture." With some reason, fundamentalists perceive a modern world of shallow, transient, secular ideologies. Pluralism or diversity, relativism, and rationalism all undermine the traditional constants and the certitude of a community guided by fundamentalist morality (Lawrence, 1995). Threatened by modern beliefs, fundamentalism has expanded in Islam, Judaism, and Christianity. Consult Armstrong's (2000) survey to appreciate how perceived threats propel the emergence of fundamentalist movements.

America is presently experiencing a revival of authoritarianism in the form of religious fundamentalism. There is an uneasy tension between those who advocate verifiable evidence as the basis for belief and action and those who see faith as necessary and even sufficient in human lives. Since faith requires absolute certainty of belief, those of faith are often eager to impose rules of their particular faith upon others who understandably would prefer to formulate their own beliefs and manage their own affairs. This ongoing conflict between faith-based dictates and evidence-based choice has ancient roots in the more general conflict between authority and skeptical inquiry. Some past disagreements between religious dogma and thoughtful inquiry turned violent. Giordano Bruno (1548-1600) was one of the few of his era to suggest publicly that life might exist elsewhere in the universe. He was burned at the stake in Rome for that insight and other indiscretions uttered while ruminating in public. In the ongoing confrontations between faith and evidence, some of the most intense public policy disputes center on human life; how it arose ages ago, how we ought to deal with life's embryonic beginnings, and what we should allow and then expect as we make that unwelcome transition from living to dying. Never underestimate the difficulties that needy egos can generate for others.

Some Problems of Western Fundamentalism

Here I emphasize how contemporary incursions of modern secular understanding have spurred retreats to fundamentalism. There are brief examples of multiple difficulties created by recent Hebrew, Christian, and Islamic Fundamentalism. As a form of blinkered wish fulfillment, Fundamentalist Faith is hunkered down in a mental bunker against reality. My criticism of fundamentalism should not be taken as criticism of religion in general.

Hebrew Fundamentalism. Scholars who study the Holy Land's ancient history believe that for some forty human generations Hebrews were repeatedly pummeled by invasions and forced evacuations, much of it recounted in the Old Testament. Equitably solving ongoing disputes in the Holy Land is as much a challenge to the human spirit as it is to reason. Jewish Fundamentalists embed their demands in the region's ancient history. One encounters raised fists brandishing non-negotiable demands empowered by the pain lingering from recent and ancient social injustices. They justify the seizure of Palestinian lands by citing passages set down in Deuteronomy centuries before Christ. This scripture instructs the Israelites to cross over Jordan (river) and dwell there. This goal is backed by 10 million zealous Christian Fundamentalists, Christian Zionists who believe that God said the Jewish people were to live in Israel. Ironically, these Christian Zionists are helpful because they are convinced that a stabilized state of Israel will clear the way for Christ's Second Coming and the disappearance of Jews by conversion or otherwise.

When a narrow interpretation of ancient scripture controls people's lives, it can blot out a meaningful hearing for humanitarian concerns. Hebrew Fundamentalism recently surfaced in their complaints about a book written by Jonathan Sacks, the chief rabbi in Britain. In essence Rabbi Sacks said that because God speaks through the scriptural writings of many religions, no one creed has a monopoly on spiritual truth. Orthodox Jews promptly condemned this reasonable view as heretical. In short order they forced Rabbi Sacks to withdraw his book from sale and rewrite it.

Christian Fundamentalism in the United States. The desire of fundamentalists to seize political power is not new. The Founding Fathers of the United States were all too familiar with the long history of strife and discord as powerful religions scurried to pull the strings of European kingdoms or civil governments. America was settled by wave after wave of religious minorities fleeing oppression by religious majorities and their powerful allies. Ironically, the desire to have one's own sect prevail made religious intolerance commonplace in the original thirteen American colonies. In each colony the dominant religious sect maneuvered to exclude others from power. Similarly, in pining for a theocracy's imagined security, present-day Christian Fundamentalists attempt to limit public education in the United States and void the theory of evolution. Some want to teach young children an untruth—that "Intelligent Design" is a scientific theory rather than religious creationism repackaged and renamed. As discussed in Chapter 15, the fancied actions of a grand Designer require an incoherent account of the evolution of complex biological systems.

Fundamentalists oversimplify life on Earth by compacting history into a more easily grasped timeframe. Many of those who have foreshortened Earth's 4.5 billion years of history into a few thousand years believe that Jesus will soon return, perhaps after an apocalypse or Armageddon. It is not unreasonable to have apocalyptic thoughts when human lives are pummeled by the threat and reality of AIDS, addictive drugs, flu pandemics, repeated warfare, and the risky behavior of nations sprouting biological, chemical, and nuclear weapons. When a natural calamity strikes, it can set already worried minds to imagining doomsday scenarios culminating in the return of Jesus Christ.

Millennialism. Conservative Christians have varied views of the circumstances of Jesus' return, depending upon their particular interpretation of a murky biblical passage, Chapter 20 in the Book of Revelation. The timing of Christ's return depends upon whether one sees human affairs as getting better or getting worse. Inclined to doom and gloom, pre-millennialists believe Christ will come to revive humans only after they have totally botched up life on Earth. Fighting and sinning on a huge scale will create such a Hellish Earth that Christ will need to return to defeat the forces of evil and give us a millennium of happiness. A century ago fundamentalists seized upon the Scofield Reference Bible (1909) because its annotations provided exactly the message they wanted to hear. Civilization was destined for destruction just as Rapture would lift favored fundamentalists safely to Heaven. Pre-millennialists believe a dictatorship of the Anti-Christ (insert the name of some leading undesirable figure here) will cause civilization to endure a terrible period of tribulation. The social deterioration we are experiencing now is the forerunner of centuries of suffering and tribulation that will end only when Jesus returns to conquer Satan at Armageddon and reign over a peaceable kingdom on Earth for 1,000 years. As a part of this odyssey, Christ is not expected to return for the grand battle with the Anti-Christ until the Jews return to the Holy Land. For this reason the Reverend Jerry Falwell viewed the day the state of Israel was formed in 1948 as second in religious importance only to the ascension of Jesus (Fundamentalist Journal of May, 1988). The re-activated Dispensationalists, who make up about a third of evangelicals in the United States, have a plan for escaping the horrors of this tribulation. In a time of Rapture, just before tribulation descends upon all non-believers, Christ will ferry the Dispensationalists to Heaven.

The more optimistic post-millennialists believe that once humans accept Christ and purify the world, he will ultimately return to preside over Heaven on Earth. If worldwide Christian evangelism somehow manages to persuade all humans to accept Christ, it will usher in a millennium of peace and prosperity capped off by the return of Jesus. Post-millennialist Christians envision Earth becoming Heaven after it has been rescued by good works. Bishop Wright (2006) is among the many who want to transform this planet until justice, spirituality, healthy relationships, and beauty prevail—a Heaven on Earth. In his view humans are presently suffering the consequences of their rebellion and corruption. Through Jesus, God wants to mount a compassionate rescue for human renewal. Wright believes one is not "snatched up into heaven" but heaven envelops Earth. He hopes that natural disasters will cease, and an abundance of harsh and difficult lifestyles of organisms will disappear as lambs lie down with lions. Wright's desire for a death-free Heaven on Earth seems particularly challenged by the realities of an expanding population, limited resources, and accumulated fears and hatreds. Nor is his timeline clear, since God's plan through Jesus to save the planet has been running for 2,000 years with agonizingly slow progress. Exactly when does the rescue kick in? Bishop Wright is largely silent on these concerns. It seems that the murkiness of Revelation 20 serves as a Rorschach-like inkblot which allows fundamentalist sects to project views of reality that fit their spiritual needs. Further accounts of the history of millennialist beliefs may be found in Armstrong (2000) and Kirsch (2006).

The vast majority of Christians in America are guided by beliefs that are neither dogmatically controlling nor callously dismissive of other viewpoints. Mainstream Protestantism has recovered from the rigidity of the Reformation and its drift toward reliance on biblical literalism. Contemporary Protestants are united more by their Christian love than by their specific doctrines. Yet, a free and open America is vulnerable to the social discontents of homebred religious militants. Like many Americans, the Christian Right is concerned about social excesses including alcohol, drugs, sex, and violence, rampant government deceit in a culture of greed, widespread corporate corruption, and other depravities. Their favored solution is to outlaw immoral behavior and install judges who will get tough with sinners—jail may be too good. Perhaps to forestall our slide toward tribulation, a group called Christian Reconstructionists wants to apply Old Testament laws requiring death for homosexuals, adulterers, prostitutes, and drug users. We are not speaking here of fundamentalists like the Amish and the Mennonites who have an isolationist strategy to avoid contamination by secular depravities. They typically skip out on public education after the 8th grade and try to remain apart from the swirl of urban sin. In contrast, the fundamentalists, or more generally the Christian Right, press their absolutist beliefs on government, on the law, on the judiciary, on public education, all in the attempt to dictate America's social values to fit their desires. In the United States the business side of faith is a multi-billion dollar industry extending from printed books and videos, to radio, TV, and the Internet, to advertising, education, finance and investment, manufacturing, real estate, and not least religious denominations and their mega-churches. In the late 1980s Jerry Falwell's Thomas Road Baptist Church of 18,000 strong raised 60 million dollars aided by broadcasts from 600 radio stations and 392 television stations. The televangelism of Pat Robertson, Jimmy and Tammy Faye Bakker, Jimmy Swaggart and others was raising millions of dollars every week. This is not your grandmother's Bible class.

Today, as fundamentalists increasingly hunger for social and political domination of America, a Christian theocratic takeover of the United States government even seems possible (Phillips, 2006). The Christian Right aims to constrain all citizens within the boundaries of its strict beliefs. Scientific rationalists who imagine that religion is a spent force in the United States are fantasizing on political, economic, and motivational fronts.

Islamic Fundamentalism. For centuries after Mohammed, Middle Eastern Islam made significant scientific explorations, embraced advances in medical practice, and encouraged the importation of agricultural products like rice, sugar cane, and spinach. By the 13th century they also had translated many foreign books into Arabic. However, matters began to change in the 16th century when Islam banned printing presses over fear the spread of diverse writings would undermine faith. (Perhaps they had reason to fear printing presses, considering how they had generated a much wider circulation for Luther's handwritten 95 theses and advanced the cause of the Protestant Reformation.) Insularity continues to be a notable feature of present-day Islam in the Middle East. It is said that Spain currently translates as many books into Spanish in one year as Islam has translated into Arabic in the past 500 years.

The Qur'an and the hadiths are considered to be the literal word of God—sacred and inviolate canons of belief. The British Museum has the oldest complete copy of the Qur'an which dates from 790 CE. There is a similarly ancient copy in the Topkapi Museum in Istanbul and another in Tashkent, Uzbekistan. These writings establish core Muslim values which include strict controls over education, written and oral expression, women's place, codes for dress and food, and the requirement for prayers five times every day. There is little interest in secular governance since, by tradition, the actions and outlook of most Muslim governments are infused with Islamic beliefs. Under orthodox Islam all aspects of life are subject to Islamic law and policy. There is no attempt to separate the state from religion—they are blended into one. This may appear odd to those whose spiritual needs do not require a politically dominant religion, but it makes perfect sense to a fundamentalist. Indeed, regardless of their particular religion, all fundamentalists believe that their canon, as the only true canon must apply to all humanity. So if you are absolutely certain that your views are the universally correct ones, it makes perfect sense to press your government to adhere precisely to those views. Fundamentalists may prefer theocracies to democracies. Like most individuals, you probably have no sympathy for those who would commit treason against your country, since it threatens your way of life. But in a theocracy where religion and government are merged, it would be political treason merely to question scripture or theological doctrines. Americans, in particular, ought to consider whether Christian Fundamentalists are on a track that parallels Islamic nations where the combination of cultural insularity and theocratic governance has slowed progress in human rights and restrained economic, scientific, and technical advances. Historically, the economic and political fortunes of Islam declined as Islamic Fundamentalism rose. Today, most Islamic economies in the Middle East have a minimal technical base; the economies float almost entirely on that remarkable dark liquid we call Oil.

In contrast with the tolerant core that comprises mainstream Christianity, the center of Islamic practice appears to embrace fundamentalism. True, a search will turn up Muslims who disagree with dictatorial theocratic governance, but it is necessary to search out such moderates—they lack power. The dismissal of outside views by a population of like-minded, often economically deprived, people can lead to the excesses of activism. Wherever fundamentalism is the majority view, it raises the risk of confrontation. If right-wing American Christians gain substantial political control, a titanic power struggle with Islam is more likely. A Fundamentalist Christian theocracy arrayed against Middle Eastern Islamic theocracies could give Osama-bin-Laden his vaunted Holy War. Sadly, the evidence-based approach to life must now struggle with Christian fundamentalists and Islamic fundamentalists poised to lock horns in a titanic conflagration.

Box 16.1 A Tragedy of Religious Zealotry.

In July of 2002 a fire broke out at the Girls' Intermediate School No. 31 in Mecca, Saudi Arabia, renowned as the holiest city of Islam. Hundreds of panicky teenagers rushed into a narrow stairwell only to find its exit door chained closed. Although firefighters arrived promptly, the muttawa fought their efforts to free the children. As the Society for the Promotion of Virtue and the Prevention of Vice, the muttawa's duties include enforcing the requirement that women wear head coverings in public. When the schoolgirls attempted to flee without headscarves, some were shut inside the burning building by the muttawa. Eventually the police subdued the muttawa and freed all of the schoolgirls, but not before 15 had died and 40 were injured. The good news is that local newspapers were able to print a vivid account of this excess in religious zealotry. Most Saudi Arabians believed this tragedy was unacceptable; the schoolgirls should have been permitted to escape. In its absolute commitment to the rigid enforcement of codes of religious conduct, the muttawa had lost sight of the value of human life.

These brief commentaries on fundamentalism in Judaism, Christianity, and Islam illustrate some of the rigidity encountered when civil society tries to reach out to fundamentalist movements to attain, if not mutual understanding, at least a peaceful détente. With mass destruction more easily accomplished today, maniacal zealotry on behalf of any religion is a significant threat to humanity. It is difficult to negotiate with those who are firmly convinced by faith they are right and others are wrong.

Animated into action by the skepticism of others, Christian, Jewish and Islamic Fundamentalisms exist as embattled faiths that gain strength when their beliefs are under siege. The more we learn about the natural world, the less we believe of the ancient myths that underlie fundamentalist doctrines. When secular facts make it difficult to deny the fanciful character of certain ancient narratives like Genesis, fundamentalists may hunker down, clutching onto their myths with growing anxiety. Sense how the secularism-fundamentalism conflict generates a mutual fear of annihilation in "...as it grows and swells, it threatens your own spiritual existence and eats away at the roots of your own world, prepared to inherit it all when you and your kind are gone...[It] threatens to destroy all that is dear and holy to us." Although this could well be a fundamentalist viewpoint, Amos Oz (1983) was actually articulating his own <u>secular</u> fears of Jewish fundamentalism during his visit to ultra-Orthodox areas of Jerusalem (quoted in Armstrong, 2000, pp. 353-354). Fundamentalisms are models of human intolerance.

A Brief Look at Evangelism

American fundamentalists aim to convert us from sin to purity whereas evangelical Christians are keener to convert us to Jesus and Salvation (Wuthnow, 2004). Fundamentalists focus on protecting their families against modern sins. Evangelists are more eager to spread the fellowship of Christ. According to George Barna's polls in the United States about a fifth of born-again Christians qualify as evangelicals eager to share their message that God's grace provides eternal Salvation. For Evangelists it can be irresistible to envelop disbelievers in the shared joy of the community of the saved. In proclaiming their faith-based Salvation, some Evangelists may exude an aura of assured, almost smug, confidence. Several motives underlie the attempt to evangelize others. In addition to spiritual conversion there are compassionate offers of food, clothing, and medical care. However, several denominations assert that arrangements between God and humans can only be set right when all infidels have been converted to the proper faith. Evangelicals may pressure disbelievers to submit to Jesus in order to fulfill their own aspirations. Naturally, they are quite convinced that the disbelievers will also benefit enormously from conversion. Yet, the Protestant reformation's goal of a 'priesthood of believers' required broad public education. As Martin Luther put it in his harsh plea for education, "I shall really go after the shameful, despicable, damnable parents who are no parents at all but despicable hogs and venomous beasts, devouring their own young." (Schultz, 1967). In contrast, it later became a common motif among evangelicals that a strong commitment to reason and learning would threaten faith. As Martyn Lloyd-Jones warned, "If you are out for intellectual respectability you will soon get into trouble in your faith." (Murray, 1990). Noll (1994, p.3-4) has noted the shallowness of evangelical education, sometimes mired in medieval conceptions: "Evangelicals sponsor dozens of theological seminaries, scores of colleges, hundreds of radio stations, and thousands of unbelievably diverse parachurch activities---but not a single research university or a single periodical devoted to in-depth interaction with modern culture." A more hopeful sign today is the increasing concern many young American evangelicals display for issues like climate change, human rights, immigration reform, and justice for the poor.

Faith and Militancy in America

Is there no hostile aspect to zealous Christian proselytizing? Consider the centrist views of the Christian evangelical movement whose professed mission is to convert non-believers to Christ in order to save their souls. Charles Marsh is a professor of religion at the University of Virginia and a self-proclaimed Evangelical

Christian. He became concerned about the stream of belligerent sermons delivered by evangelical ministers in the run-up to the second Iraq war in 2003. Instead of displaying forbearance and a longing for peace, evangelical leaders rattled their sabers at Iraq. In one sermon, Charles Stanley, a former president of the Southern Baptist Convention exclaimed, "We should offer to serve the war effort in any way possible. God battles with people who oppose him, who fight against him and his [Christian] followers." Franklin Graham, the son of Billy Graham, enthused about a possible American invasion of Iraq that would create "exciting new prospects" for proselytizing Muslims. The question of "Whom would Jesus kill?" seemed of little concern to the stunning 87% of white evangelical Christians who supported President Bush's decision in April of 2003 to invade Iraq preemptively. Jerry Falwell boasted in the title of his 2004 essay that, "God is Pro-War." What did Professor Marsh make of Christian love transformed into belligerence reminiscent of earlier Christian Crusades? "The single common theme among the war sermons appeared to be this: our president is a real brother in Christ, and because he has discerned that God's will is for our nation to be at war against Iraq, we shall gloriously comply (Marsh, 2005)." This is a volatile mix of arrogance and authoritarianism inflamed by faith. We need to hold Americans to a higher standard.

Faith and Absolutism

Whether it is a secular or a theocratic dictatorship, absolutism guarantees oppression. You can rely on authoritarian regimes to deploy fear, intimidation, and torture while defending their power and ignoring pleas for power sharing. Beware of faith wearing a sword. The Qur'an, like the Old Testament, is full of God's wrath, vengeance, and all-around general contempt for the despicable infidels; that is, anyone who doesn't accept Islam.

Many folks say they are seeking faith. After all, what harm can be done by having faith. If your faith is correct, then you are on your way to Heaven, and if it is wrong, you are no worse off for the effort—hang the problems you cause by imposing your faith on others. Why is there a longing for absolute faith, if science is able to make accurate predictions and create things that work with no recourse to absolute certainty? It should be possible to make workable religious commitments without claiming absolute truths that will hold for all time. Some moral attitudes and beliefs portrayed as absolutely correct in earlier times have changed so markedly through the history of Christianity that neither the devout nor any one else accepts them now. Once widely accepted, slavery now repels, as does burning witches and children at the stake, the forced exorcism of Satanic infestations, and torture and dismemberment as punishment for heretical views. True Christian faith offers loving advice and guidance; it does not impose on the innocent the current version of absolute dogma. Does God want us to get things right or to have blind faith? It can't be both ways.

Social impacts of faith. If faith trumps evidence, might it stall science and replace facts with supernatural myths of pre-scientific eras? A theocracy based on faith is the fond hope of some. Their vision would see the disappearance of the evils of: drugs, Hollywood's debauchery, the liberals and their diversity, science that imperils religion, rap music, and the iniquity of inner cities. Paradoxically, the science-dependent gadget-filled suburban house close to hospitals brimming with medical advances would somehow remain. There is a yearning to have a simple morality and a stable dogma instead of the modern culture of skepticism and ceaseless inquiry. As two neatly attired Mormon missionaries put it to me, "College education destroys faith." Or as I might put it, "Faith can't withstand scrutiny by the informed mind."

One could cite thousands of examples in which the absolutism of faith has provided an easy rationalization for evil actions, including wholesale slaughter. The plight of the Cathars in southern France is a classic example. By claiming that God spoke to each person directly, rather than through the priests or scripture, the Cathars made the priests seem superfluous. In denying the authority and hierarchy of the institutional church, the Cathars also dodged the imposition of tithes, penances, strictures, and obsequies, not to mention baptism and communion. This insubordination was too much. In 1209 Pope Innocent III directed 30,000 Knights Templar and their foot soldiers to suppress the Cathar heresy as a threat to the preeminence of Rome. A wave of death and bloodshed engulfed town after town, literally burying Cathar views while preserving the institutional church. In the town of Beziers alone, more than fifteen thousand men, women, and children were slaughtered—strictly for their own Salvation, it was claimed. In this the papacy rationalized, as it always has, that the afterlife had a greater value than mere earthly life. Human claims about the likely fate of your soul trump your imagined right to life on Earth. When one officer of the Knights Templar inquired as to how he was to distinguish true believers from heretics, the papal representative is reputed to have said: "Kill them all. God will recognize his own." Beyond the loss of thousands of lives, the monumental loss for civilization was to relinquish the most cultured region of Europe to the barbarians.

Personal deviations under faith. Fundamentalism is a fantasy life that can spawn selfish, even dangerous or corrupt behaviors by those who believe that regardless of their misbehaviors, they will be saved by Jesus Christ. Being saved by faith alone provides wonderful opportunities for narcissism. Not only am I free to do what I want, but also from time to time I can disregard my commitments to other human beings because I am Heaven-bound. I can walk away from social responsibilities by claiming a higher responsibility—obedience to the higher authority of God's will. How do you know you are following God's will? Basically, if you strongly sense it is God's will, it must be God's will. Employers must judge whether they are hiring a wonderfully sincere and loving individual of faith or a model of narcissism, responsible only to the individual they see in their mirror. Pretending that personal motives are God's will is not simply the common man's evasion. It has been an especially useful ploy for kings, dictators, presidents, and other major leaders who have frequently claimed divine authority in support of their self-serving actions.

Current examples are not hard to find. Frank Turner was a capable and engaging anchor for Channel 4's news programs in Detroit until his raging cocaine addiction eventually ruined his professional and private lives. Two divorces and two bankruptcies later he kicked his drug habit with the help of God—quite a commendable feat with a drug as controlling as cocaine. Having re-made himself into an evangelical minister, Turner booms his inspired testimony of grace and redemption. In his own resounding words on November 21, 2006, "The only debt I owe is to God and that has been fully paid [by being born again]. I don't owe anyone else anything." Actually several people, who are each still out more than $10,000, would like to be repaid, but Frank Turner has already settled up his account with God.

Some believe that it is a sufficient justification to point to having good intentions in trying to follow God's will. This motive may provide immunity against criticism in some religious denominations, but a defense of good intentions in medicine surely won't impress when an obviously better alternative was not used. Consider the case of the hapless medical student diligently caring for a patient with digestive problems related to a shortage of intestinal bacteria. As a medical remedy, the student served up a chocolate milkshake secretly dosed with homogenized fecal material! The intentions were good, and the bacteria may actually have cured the patient—but yuk. Please don't wing it, if bolstered only by the purity of your intentions.

People reveal themselves both by what they embrace and by what they attack. Why do people of faith so frequently attack humanism? Humanism is not some bizarre cult with strange and secret observances; it is an open and prudent stance that affirms love and peace on Earth. So where is the threat in this? It is not because humanists have laxer morals or are more inclined to a survival-of-the-fittest approach to life. Rather it is the humanists' principled refusal to commit their beliefs to the supernatural and their souls to evangelical entreaties that sends shivers of uncertainty or pity down the spines of the faithful.

Box 16.2 King of the Hill.

Tribal hatreds, contested land claims, and the disruption of sacred sites in Jerusalem have provoked some of the world's most intense religious feelings. According to some traditions, nearly four thousand years ago the covenant between Abraham and God set aside lands in Palestine for the tribes of Israel. It was there in Jerusalem that King Solomon built his defining Temple. During Roman times the city of Jerusalem was the seat of Israeli power until Solomon's already rebuilt temple was destroyed yet again in 70 CE. For Islam the city of Jerusalem is the third holiest, outranked only by Mecca and Medina. It has been under nearly continuous Muslim control since the time of Mohammed (638 CE). Rival Jewish and Muslim claims for Jerusalem come into sharpest focus at its Temple Mount. In 691 CE a major Muslim monument, the golden Dome of the Rock, was erected on Jerusalem's Temple Mount as the site where Abraham reputedly offered his son to God for sacrifice. Unfortunately, this is also the site of the remnants of Solomon's reconstructed temple. Portions of its Western Wall have become the holiest site in Judaism. The traditional Jewish view has been that Solomon's temple will be rebuilt only after the Messiah comes. However, in the 1980s Jewish Fundamentalists began to imagine that if they rebuilt the Temple first, it might attract the Messiah. Some extremists even believed the Messiah would surely come if they blew up the Islamic Dome of the Rock on top of the Jewish Temple Mount. Fortunately, final preparations to destroy the Dome were cancelled when no rabbi could be found to approve such a horrific plan. Given the multiple interlocking international treaties and religious alliances, some believe destruction of the Dome of the Rock could have precipitated World War III (Armstrong, 2000). Like Christians, who razed pagan temples for sites to build imposing churches, the Muslim's earlier construction of the Dome of the Rock was doubtless intended to overwhelm Jewish competition for the Temple Mount. Human anguish continues over these competing sacred symbols—dual dreams for rock and stone venerated to the extreme.

Testing the Strength of Faith

It is natural enough to want to avoid death. Most of us would be delighted at the prospect of an everlasting life. Every Sunday many of us stand shoulder to shoulder in ritual solidarity stiffening as best we can our fragile faith in everlasting life. Entry into Heaven is our expectation because we have been saved by faith in Jesus as our personal Savior. But how strong is this conviction? Religion's promise of everlasting life in Heaven requires a mind-warping leap of faith to be convinced it will literally deliver us into that distant Promised Land. Countless Americans have attempted leaps of faith. They really want to believe, but know it is delusional. Like hypnosis, faith works best for those most open to suggestion. Perhaps Heaven on Earth?

Dire circumstances can provide a good test of the strength of faith. Suppose Dwight, a man of faith, slips while climbing the face of a mountain. As he begins to slide down toward certain death, Dwight realizes he must quickly choose one of two options. Either he grabs a nearby plant scientifically demonstrated to have a deep and sturdy taproot, or he grabs a nearby blanket of sacred moss which his religious faith certifies will protect true believers from harm. Do you think Dwight should grasp for the deep roots or the sacred moss?

In another setting an elderly man says, "My religion provides me with a complete and abiding faith. In just six months I will finally be released from the snake eyes of my terminal bone cancer and its excruciating agony. At last I will most assuredly enter the kingdom of Heaven to exchange this wretchedness for eternal happiness and bliss." What is wrong with his reasoning? If his faith is so strong and the pain so intense, why not exit Earth for eternity now, what is the eyeblink of a few months against the forever of eternity? Self-professed believers may nonetheless desperately cling to their own earthly lives.

So we need to ask why some, who seem to be among the most convinced they will enjoy everlasting bliss in Heaven, are also among the most reluctant to leave Earth. Not only do they want to prolong their existence on Earth as long as heroic medical interventions will permit, but they may oppose a stranger's attempts to "die with dignity." The answer may lie in evolution—we are creatures determined to cling to life. Today a person truly terrified of death is being consistent by adopting a faith that life will continue in Heaven while simultaneously stifling the actions of others seen as potentially shortening his life while on Earth. The opponents of death-with-dignity laws are not animated by an abstract veneration of life. The typical opposition is an intensely self-serving attempt to lessen their fear that someone might kill them when they are helpless. Sad, but understandable.

In the process of tightly clutching their religious beliefs, the insecure may shower unbelievers with threats and hostility. In responding to the ire of a fundamentalist preacher the skeptic may say, "Just because I do not wish to comply with your customs, why am I hated as if I were despicable?" This real confrontation was actually an early Christian retorting to the Romans who were pressing their belief in multiple gods upon the Christian population. At that time Roman authorities feared that Christian beliefs might reduce the power of the Roman government.

The Authoritarian Personality

An individual with an authoritarian personality (one who needs to obey the dictates of powerful authority figures) may find satisfaction in a life of obedience to the Highest Authority. Throughout history, obedience to the head person has structured life for the subservient. Hierarchy has been a repeated theme: braves serving the chief, scribes serving the pharaoh, privates obeying their captain, citizens obeying the Roman emperor, priests obeying their bishop, vassals serving their lords, samurai serving the shogun, a worker serving his boss, an assistant serving the professor, and the devout serving God. In Japan the Samurai's code of loyalty fed into the "victory or death" outlook of Japanese soldiers in World War II, most dramatically in the kamikaze pilots. In earlier times monks and nuns, cloistered in monasteries separated from the temptations of secular world, took vows of poverty, chastity, and obedience. These austere religious orders included the Benedictine, Carthusian, and Cistercian monastic groups of obedient monks. The mendicant orders like Augustinians, Capuchins, Carmelites, Dominican, and Franciscans were so committed to impoverishment that they relied on begging until it became unsustainable. In Antigua, Guatemala one can walk through the semi-intact ruins of a Capuchin nunnery destroyed by the powerful earthquake of 1773. Those novitiates who took the vows and adopted the cloth habits were accepted into a state of permanent silence, isolated from the outside world and largely from one another. During Lent, each penitent nun would retreat to one of the eighteen 4x8 foot stone rooms to live in total isolation for 40 days and nights. It is difficult today even to imagine this degree of isolation and privation, short of punitive high security prisons. Yet, even now obedience to God and three ascetic elements—silence, isolation, and prayer—dominate the lives of Capuchin nuns living in an active convent in Guatemala City in the 21st century. Worldly denials are offered up as proof of devotion to God.

If I kneel to authority and adopt unproven Christian credos on faith, does that bring me into conflict with secular society? Not necessarily. It may not be as exhilarating, but it is possible to compartmentalize your natural and supernatural belief-sets. However, it is another matter if one presses fantasies on others against their desires, needs, or safety. For some, the need to dominate makes religious beliefs preeminent over laws. An enormous amount of evil continues to be done through indisputably personal decisions falsely claimed to be contracts co-signed by God. That is the potential cost, the real risk of substituting the mirage of instructions from God for the reality of evidence. Evidence-based conclusions cannot be, and must not be, waived off and dismissed in order to surge forward to personal emotional fulfillment in faith.

Box 16.3 Should Our Leaders Be Loyal to This World or the Next?

"Reclaiming America for Christ" is the title of an annual conference in Fort Lauderdale, Florida which attracts Christian nationalists eager to "take back the land." Pam Stenzel, a graduate of Jerry Falwell's Liberty University, spoke about her job of promoting sexual abstinence education. Stenzel recounted her recent airplane conversation with a quite skeptical businessman.

"What he's asking," she said, " is 'Does it work?' You know what: Doesn't matter. Cause guess what. My job is not to keep teenagers from having sex. The public schools' job should not be to keep teens from having sex."

Then her voice rose and turned angry as she shouted, "Our job should be to tell kids the truth! People of God," she cried, "can I beg you to commit yourself to truth, not what works! To Truth! I don't care if it works, because at the end of the day I'm not answering to you, I'm answering to God!"

Later in the same talk, she explained further why what "works" isn't what's important—and gave some insight into what she means by "truth." "Let me tell you something, people of God, that is radical, and I can only say it here," she said. "AIDS is not the enemy, HPV and hysterectomy at twenty is not the enemy. An unplanned pregnancy is not the enemy. My child believing that they can shake their fist in the face of a Holy God and sin without consequence, and my child spending eternity separated from God, is the enemy. I will not teach my child that they can sin safely" (Goldberg, 2007, see p. 135).

Appointed by President Bush, Pam Stenzel serves on the task force at the Department of Health and Human Services to help implement abstinence education guidelines. She speaks to half a million children a year as part of an 88 million dollar a year abstinence program. In April of 2007 a commissioned, careful analysis concluded that this abstinence program has had no significant effect on US teenagers' sexual activity (www.mathematica-mpr.com).

Does Your Plan Rely on Faith or Good Works?

Spirituality wells up in those Christians seeking something greater than themselves. The actual manifestations are reflected in major differences among Christian philosophies. Some wings of Christianity focus on accessing Heaven above. Other Christians work to ensure justice for all, and generally try to set things right on Earth rather than seek a remote Heavenly world. Christian monks may prefer detachment to involvement with human affairs.

Those Christians who want an accurate picture should merge three sets of information into a coherent framework. First, one tries to extract relevant historical truth from the Bible and other records about the period surrounding Jesus. There is no shortage of debate about the accuracy of such descriptions. Second, the reading of biblical history and other sources must not be blatantly inconsistent with today's substantial scientific understanding of the universe and life on Earth. Third, contemplation of the purpose and meaning of one's life should result in a plan.

Paul realized Christianity could not be subservient to the Torah and the existing laws of the Holy Land. "…for if righteousness comes by the law, then Christ died in vain" (*Galatians.* 2:21). In other words, if you can be redeemed or justified merely by being law abiding, then who needs Jesus? This is one reason that Paul considered redemption less dependent upon performing good deeds than committing to a faith in Jesus. However, Paul also made it clear that Christians were to "uphold the law" and not void it through faith (*Romans* 3:28-31). Faith becomes a societal menace when secular laws are superseded by doctrinaire values, based upon the self-serving view that they are impelled to do God's work.

Can you be saved merely by professing Christianity? How much must one contribute to Christian service as indicated in the Epistle of James? And in *John* 5:28-29 "The hour is coming, in which all that are in the graves shall hear his voice, and shall come forth; they that have done good, unto the resurrection of life, and they that have done evil, unto the resurrection of damnation." [When Jesus returns, the past deeds of the revived dead will merit either Heaven or Hell.] Christians continue to be divided on the necessity of good works or the sufficiency of faith. This fracture cannot be healed by the observation that those with proper faith will, as a matter of course, automatically do good works in Jesus' name. What about those who can't muster the needed faith; will recourse to good works be sufficient for their Salvation? Who believes there would be a multitude of denominational viewpoints on Salvation if the rules for accessing Heaven were clear? Every Christian should make their personal examination of the balance between faith and good works. Institutional religions can claim many good works in the world of human affairs.

17

&

Some Contributions of Christianity Today

Moral Progress and Religion

Gaining voting rights for women, increased sensitivity to sentient animals, diminished interest in blood sports, an increased revulsion over the carnage of war, and the renunciation of slavery all suggest continued moral progress. Many of these reflect better scientific understanding about life and greater awareness of pockets of abuse that were once hidden from public view. Not that there aren't numerous troubling issues such as frequent wars, abuse of drugs, gambling, and other kinds of self-indulgence.

Some argue fervently that religion has outlived its usefulness and should be abandoned wholesale and retail (Harris, 2004; Dawkins, 2006; Hitchens, 2007). These critics focus on the excesses of faith and ignore the contributions of religiousness. In arguing against the perils of blind faith and touting the importance of an evidence-based assessment, the present book has necessarily pointed out historical religious abuses. Does this mean that religion can point to no achievements? Clearly not. Institutional religion has made substantial contributions both in the past and in the present, some of which I will highlight. And many thousands of mainstream Christians are courageously working in their own denominations to raise those in control of institutional religions to a higher moral plane.

Institutional Religion has Founded Colleges and Universities

One way to appreciate the value of institutional religion is to imagine American society today in the absence of early Christian influence. Certainly the public would be less well-educated and informed. As romantically heroic as their achievements might have been, few frontiersmen, cowboys, and homesteaders had the inclination or the extra energy to found colleges and universities in America's westward expansion. Today the United States has approximately 3,000 universities and colleges. Of the 1,600 private institutions, 900 hundred remain church related, of which 220 are Catholic institutions like Notre Dame and Villanova Universities. Historically, virtually all major private universities and many public institutions were founded through sectarian church support, even if the affiliation is scarcely apparent now. The Congregational Church founded 45 colleges and universities including Harvard in 1636 and Yale in 1701. A quick sampling of other large and small institutions of higher learning founded by Protestant denominations includes Wake Forest (Baptist), Lafayette (Presbyterian), Gettysburg (Lutheran), Ohio Wesleyan (Methodist) and Bryn Mawr, Haverford, and Swarthmore Colleges (Quaker). Americans owe institutional religion an enormous cultural debt, even as we might take exception to the stultifying education offered by some theocratic college administrations.

When religious denominations pushed for education and inquiry, they advanced medicine and engineering, and research. It helped to make America a leader in higher education. I for one am extraordinarily grateful to these forward-looking Christians who founded colleges and universities instead of spending these same resources to build churches all the bigger to save additional skeptical souls. Education was a higher calling. My father, who taught briefly at Villanova University after retirement, was aware of the low salary scale for the Catholic priests who had been called to teaching. One priest reflected on how he managed with the rising cost of living. "Yes, it is true," he said, looking Heavenward, "my salary is modest, but ah, the fringe benefits!"

Charitable and Humanitarian Contributions of Religion

In addition to the crucial role of institutional religion in establishing some pillars of American higher education, we should not neglect religion's core mission. By providing a feeling of hope, which lifts and comforts human hearts, the spiritual benefits to individuals are tangible enough. Reassurance and a sense of redemption sustain many who cannot abide by naturalism's worldview. Lying beyond the spiritual impact upon individuals is the immense scope of charitable works and humanitarian responses that benefit society at large. In addition to the actions of governments and private foundations that address human needs, an enormous range of social support also flows from religious sponsorship. It seems that the protective sense of humans frequently urges them to come to the defense of the vulnerable. The infirm, the sick, the old, the frail, the blind, halt and lame, the young, the infant, and yes the fetus, are all vulnerable. In addition to religious mandates to do good works, concern for the vulnerable may be felt because we can imagine ourselves or our kin in similar predicaments. Charity may be an aspect of reciprocal kindness (Bowles, 2006; Nowak, 2006) and parenting circuitry in the brain. Whatever the reasons, the urge to comfort the vulnerable is part of our human makeup and we need to listen to it.

In their capacity as resources serving society, thousands of charitable religious groups actively contribute to human welfare by stepping in to relieve human suffering. In 2003 the Internal Revenue Service identified 937,000 public charities in the United States of which more than a third were religious and the remainder secular. Religious and secular groups each received about two-thirds of their funding from government sources. The social services provided by organized religion include such diverse programs as:

Aid to the hungry
Assistance for newcomers to a community
Assistance to shut-ins and the aged
Childcare
Civil rights
Counseling for individuals
Disaster relief
Family counseling
Home construction and repair
Literacy and tutoring programs
Medical services including health care and prostheses
Orphanages
Schools
Second hand clothing and used household goods
Shelters for wayward children and broken families
Soup kitchens and warm shelters for the homeless
Religious youth groups that assist the urban and rural needy

These varied projects of religious charities provide remarkable improvements in people's lives. Many churches have a policy of outreach to spend a set percentage of their resources on community needs. Generally these local projects like housing, or literacy, or care of the needy fit well with traditional social concerns. Local, small-scale community projects are better able to provide compassionate understanding of personal circumstances. There are thousands of charitable religious groups, some well-known, others working in obscurity, but all trying to contribute to human welfare.

Some sense of the scope of community services offered by United States Christian congregations is provided by a 1988 survey that obtained responses from 1353 of 4205 notified congregations (Hodgkinson, et al., 1988). A high percentage of responding churches engaged in family counseling (79%) and youth programs (79%), while also supporting a wide range of other programs including: community development (46%), support for hospitals, clinics, nursing homes, and hospices (44%), civil rights and social justice programs (42%), meal services (38%), aid to refugees (35%), shelter for the homeless (32%), day care (31%), after-school programs (30%), teenage pregnancy counseling (29%), tutoring (26%), battered women programs (25%), and housing for seniors (19%). Larger congregations (greater than 400 members) were more active socially. An average of 22% of individual members participated in these social programs with a typical congregation allocating 5% of its total budget.

Box 17.1 A Sample of Christian Service.

"John Lee is a Chinese American in his late twenties who works as a teacher and attends church services every week at a large suburban Presbyterian church in eastern Pennsylvania. Through the church he has made close friends with several other people and has gone on church-sponsored trips with them to China and Nicaragua. Several years ago he found himself with time on his hands. His friends at the church encouraged him to talk to the pastor of a large inner-city church that sponsored service projects. Mr. Lee had enjoyed his previous international trips, so he was delighted to learn that the church was sending a team of lay volunteers to the Dominican Republic for a week to do repair work at an orphanage. He spent the week there, putting screens on windows, cleaning, painting, and getting to know the staff and children. He has now made three of these annual trips." Mr. Lee says he is motivated both by his desire to follow in the footsteps of Jesus and also by his enjoyment of international travel. (Excerpted from *Saving America?*, Robert Wuthnow, 2004).

In addition to meeting ongoing needs through aid to the hungry, homeless, aged, orphaned, sick, addicted, or otherwise disadvantaged, many religious charities target disaster relief as a major humanitarian concern. Fires, floods, hurricanes, earthquakes, and disease epidemics are some of the large-scale disasters targeted by religious charities. To help with the December 26, 2004 Indonesian tsunami, $150 million in aid was provided by religious groups which included: Adventists, Quakers, Roman Catholic, Orthodox Christian, Baptists, Mormons, Lutherans, and various other Jewish, Christian, and Islamic groups. About half of the groups providing aid for tsunami relief were non-religious. The Catholic Charities USA Disaster Response brings together social service agencies operated by most of the 175 dioceses in the United States. The Christian Disaster Response cooperates with the American Red Cross, the Salvation Army, Church World Service Disaster Response, and NOVCE to make disaster assessments and provide feeding facilities and relief supplies. It stockpiles donated goods at regional distribution centers.

The United Way/United Fund is among the best-known coordinators of contributions to charitable organizations. With an annual fund drive it solicits support for dozens of religious and secular organizations. The Salvation Army is one of the largest faith-based organizations. In the United States it operates 1369 "corps community centers," 1,640 thrift stores, 571 group homes or temporary housing facilities, 228 day care centers, 222 senior citizen centers, and 163 rehabilitation centers. It considers itself to be "an evangelical part of the universal Christian church." It believes the Bible was divinely inspired and that our Salvation depends upon the divine grace of Jesus Christ.

The American Friends Service Committee (Quakers) provides educational and social assistance worldwide. Catholic charities worldwide had $2 billion revenues in 2004 and spent 90% for programs and services; 1400 local Catholic agencies served 6.5 million people. The Christian Reformed World Relief Committee assists stricken families, repairs homes, and helps with clean up and child care. The Episcopal Church Presiding Bishop's Fund for World Relief provides grants for food, water, medical assistance and financial aid for the first three months after a disaster.

Educational Concerns for Hunger provides training to missionaries and development workers worldwide in addition to offering information about tropical agriculture and seeds. Marine Reach International provides medical care and facilities via ship and truck. Bright Hope International works through local churches in 42 countries to provide food, jobs, and aid to the needy. Heal the Nations is a Christian nonprofit organization dedicated to facilitating community health development for needy people in remote areas of the world. In addition to Christian charitable organizations highlighted in these few examples, there are dozens of charities sponsored by Buddhist, Islamic, Jewish, and other religions. It is all quite impressive.

Many Christian conservatives believe their "religion is the one true faith leading to eternal life" and that "a person cannot be a good American if he or she does not have a religious faith." They also believe that health care, literacy, and job training are better left to government agencies, whereas religious ministry is more effective in mentoring young people, treating drug and alcohol addiction, and counseling teens about pregnancy (Pew Religion and Public Life Survey, 2001).

With more than a third of a million religious charities in the United States, there is ample scope to supplement the work of government social agencies. The compassion of dedicated religious volunteers can shine through, sometimes contrasting with the wearied indifference of overburdened state and federal social services. Local congregations ministering to local needs are viewed as the most effective with a grade of A-. Public welfare departments are rated the least effective with a grade of C+. They may be rated low because government sites are not close at hand, offer limited options, and have an overworked staff. Surveyed in 2002, faith-based organizations were perceived to be no more effective than non-governmental secular organizations; both received a grade of B (Wuthnow, 2004 p. 208).

Christians, especially biblical literalists, have traditionally opted for the dominion of mankind over nature, effectively encouraging the unsustainable exploitation of natural resources (Greeley, 1993). As the movement for a sustainable environment has gained momentum, Christian denominations have shown more interest in tackling various environmental concerns. Pope John Paul's comments may be found on the web (www.conservation.catholic.org/declaration).

It is scarcely surprising that, while providing charitable aid, religious organizations busily encourage recipients to embrace favored religious beliefs. Thirty eight percent of faith-based non-profit organizations believe it is very important to transform the client spiritually in the process of helping. As long as getting aid doesn't require a religious commitment, it is not unreasonable that adult recipients be asked to listen to brief non-bullying evangelical overtures. However, I would argue against indoctrinating needy, recipient children when they are young and trusting. This is not an abstract concern since one-third of child-care agencies are church based. Those with absolutist views are likely to deny any obligation to protect children from indoctrination since they have no doubt that the indoctrination they impose is exactly what children need. Others believe that child indoctrination brands the defenseless mentally and emotionally. The increased funding of religious charities by public taxes has the follow-on effects of greater proselytizing and religious indoctrination—both are matters for deep public discussion in the context of America's founding.

18

&

Pruning the Religious Roots of America?

Re-jiggering Constitutional History

Here we turn our attention to ongoing attempts to convert a democratic America into a theocracy. If that misguided objective succeeded, it would increase the risk of a religion-driven epic conflagration with Islam. Editorial writers, talk-show hosts, clergy, and politicians energetically reassure us that America was founded as a Judeo-Christian nation. Since there were almost no Jews in America at the time of its founding, perhaps these pundits mean to refer to a nation based on Judeo-Christian values. Christian values certainly have deep roots in Judaism. Counterpart passages in the Hebrew Bible and the Christian Old Testament recount many value-laden sagas of Jewish tribes. Christianity began as an offshoot of Judaism with the teachings of a Jew named Jesus who preached to Jewish tribes in Palestine.

In America in 1777 there were perhaps fewer than one thousand Jews and only twenty thousand Roman Catholics out of 4.5 million mainly Protestant Americans. In the 1830s and 40s waves of immigration swelled the number of Catholics to 1.6 million. In reaction, longstanding Protestant mistrust of Catholics welled up in social, political, and sometimes violent anti-Catholic opposition. Clearly, "Protestant-Christian American colonies" would be a more accurate characterization of the theological roots of the early colonists. Was Protestantism then embedded in the Constitution?

As historians have noted "It is not true that the founders designed a Christian commonwealth, which was then eroded by secular humanists and liberals; the reverse is true. The framers erected a godless federal constitutional structure which was [only later] undermined as God entered first [sic] the U.S. currency in 1863..."(Kramnick and Moore, 1996, p. 143). Examples of early Protestant sectarian antagonisms in the governance of the thirteen original colonies were certainly evident to the Founding Fathers. Many of the first American colonists were English religious dissidents. We frequently laud them for their pluck and courage in leaving their homelands and making an arduous crossing of the Atlantic Ocean to escape religious intolerance in Europe. Hunkered down in small groups they understandably, yet intolerantly, demanded religious conformity in their new communities. The Massachusetts Bay Colony's actions give us some sense of the reach of intolerance in our Christian heritage. Every citizen of Massachusetts was expected to attend Congregational Church services. Quakers were so despised that by colonial law any of that "cursed sect of heretics" who entered Massachusetts were to be jailed, whipped, and expelled. The truly unlucky ones were branded and expelled with missing ears. William Penn and his Quaker followers fled south to found Pennsylvania or "Penn's Woods." The banished Roger Williams established a haven for Quakers and Baptists in Rhode Island. Quakers and Catholics living in Virginia and Protestants in Maryland were not eligible for public office. And so it went. This remains as a sad lesson for today. Even though the victims of intolerance may rise to power, they may still fail to see their way to extending tolerance toward others.

Here is the view of the Supreme Court on earlier religious intolerance in colonial America: "Catholics found themselves hounded and proscribed because of their faith. Quakers who followed their conscience went to jail; Baptists were particularly obnoxious to certain dominant Protestant sects; men and women of varied faiths who happened to be in a minority in a particular locality were persecuted because they steadfastly persisted in worshipping God only as their own consciences dictated. All of these dissenters were compelled to pay tithes and taxes to support government-sponsored churches whose ministers preached inflammatory sermons designed to strengthen and consolidate the established faith by generating a burning hatred against dissenters." (Everson v. Board of Education, 1947, 333 U.S. 1). The last sentence is worth re-reading in the context of the political impact of today's faith-based initiatives.

It is likely that the Founding Fathers were unsettled by the religious intolerance of colonial America; they may have seen enough. In each colony the largest Christian group had attempted to seize power. Already by 1787 nine of thirteen original American colonies were controlled by sectarian religious monopolies. However, with no single denomination large enough to gain control at the national level, the diverse representatives to the Constitutional Congress Convention voted against either a religious test for office or tax support of religions, in part to shield the national government from a future takeover by the eventually dominant religious sect.

What religious values were embedded in the founding of the United States? America's Founding Fathers are often portrayed as devout Christians intent on creating a Christian nation. In longing for an uplifting tale, it is tempting to tether Christianity to patriotism and imagine the Founding Fathers as "Christian apostles in kneebritches" (Meacham, 2006). Is this allowing our romantic imaginations to subvert history?

To check our desires against American history, let's begin with the three main documents identified with the Founding Fathers. *The Declaration of Independence* (1776) was signed by 56 white males. *The Articles of Confederation* (1777) was signed by 48 white males, representing the thirteen states. *The United States Constitution* (1787) was signed by 39 of the 55 white male delegates. Only two men signed all three documents. None of these documents contains Christian images or references to Christianity in spite of ample opportunity to insert such provisions.

Moreover, Article VI, paragraph 3 of the United States Constitution states that, "...no religious Test shall ever be required as a Qualification to any Office or public Trust under the United States." To establish the more general point of independence from religion, the First Amendment to the U.S. Constitution begins with an establishment clause and a free exercise clause which state, "Congress shall make no laws respecting an establishment of religion, or prohibiting the free exercise thereof." Here the framer's main concern was that public taxes not be used to support religion. Alternative constitutional language that no religion should be favored over others was rejected in the delegates' debates out of concern that all religions might then have an equal and substantial claim to be supported by public taxes. Their intention was not to curtail religious expression but to exclude government support and funding of any and all religions.

Well, didn't Christianity pervade the atmosphere of the deliberations that created these neutral documents? Wasn't Christian piety in evidence? The lengthy negotiations over the Constitution were exhausting. So much so, that after nearly five weeks of debate, a rankled Ben Franklin suggested that the Convention should begin each day with a prayer that might "illuminate our understandings," noting that ... "the longer I live the more convincing proofs I see of this truth—that God governs the affairs of men." However, the Convention, which had not prayed up to this point, promptly voted down Franklin's motion to hold prayers (Bowen, 1966, p. 125). And it was not by oversight, but after ample debate, that the Convention decided to exclude any mention of God and Christ in the Constitution.

Of course, the absence of references to Christianity in the Constitution itself leaves it unclear whether the wording of congressional legislation referred to America as a Christian nation. An early document from the Senate made an unequivocal declaration on this point. In the 1797 Treaty of Peace and Friendship with Tripoli, Article XI begins bluntly: "As the government of the United States of America is not in any sense founded on the Christian Religion..." (Bevans, 1974).

Well, if early congressional documents are unsupportive of Christian American governance, perhaps we can find support in the personal feelings and beliefs of the most prominent of the Founding Fathers. Naturally, when speaking in public, politicians are likely to pander to public sentiment, offering supportive words for God and religion. So let's peer into the private views of the Founding Fathers, which are likely to be more forthright.
In the words of Ben Franklin,

> I have found Christian dogma unintelligible. Early in life I absented myself from Christian assemblies. (In: *Toward the Mystery*, undated)

In the spring of 1790 Franklin wrote to the president of Yale University,

> I believe in one God, creator of the Universe.... As to Jesus of Nazareth, ... I think the System of Morals and his Religion, as he left them to us, the best the world ever saw or is likely to see; but ...I have, with most of the present Dissenters in England, some Doubts as to

> his divinity;...I think it needless to busy myself with [studying] it now, when I expect soon an Opportunity of knowing the Truth with less Trouble... (van Doren, 1938, p. 777)

Apparently Franklin was wryly commenting on his own failing health, for in less than two months he was dead (April 17, 1790).

In the words of James Madison,

> Religious bondage shackles and debilitates the mind and unfits it for every noble enterprise... During almost 15 centuries has the legal establishment of Christianity been on trial. What have been its fruits? More or less, in all places, pride and indolence in the clergy; ignorance and servility in laity; in both, superstition, bigotry, and persecution. (In a letter to William Bradford, Jr. April 1, 1774, as cited in Madison's famous 1785 *Memorial and Remonstrance*)

John Adams, the second President, rejected the Trinity and the deity of Christ (Nettlehorst, 2006). In his words,

> Consider what calamities that engine of grief [the Cross] has produced! (In a letter from John Adams to Thomas Jefferson, undated)

Thomas Jefferson famously took a razor to his own Bible to cut out all references on the Trinity, the virgin birth, bodily resurrection, angels, and miracles, including the resurrection of Jesus. The "Jefferson Bible" was the result. As he said,

> I am a sect by myself, as far as I know. It [the Trinity] is the mere Abracadabra of the mountebanks, calling themselves priests of Jesus...no man has a distinct idea of the trinity.

Thomas Paine had a similar view in saying,

> I do not believe in the creed professed by the Jewish Church, by the Roman Church, by the Greek Church, by the Turkish Church, by the Protestant Church, nor by any church that I know of. My own mind is my own

religion... (From *The Age of Reason*)

Politicians habitually curry favor with insincere religious pleasantries, so you have to imagine that such strongly critical remarks must have been sincere. In short, "deist" rather than "Christian" best characterizes the personal religious beliefs of the most renowned Founding Fathers (John Adams, Benjamin Franklin, Thomas Jefferson, James Madison, Thomas Paine, and George Washington.) In deism, God created the universe and established the natural laws that governed matter and energy, but thereafter rarely, if ever, intervened on Earth (but see Franklin above).

Theocracy in the Making?

Apprehensive about any competing authority, heads of nations have historically asserted their full power to control religious matters. Reminiscent of the sweeping powers of the recent Patriot Act in the United States, King Henry VIII's signature on the Act of Supremacy in 1534 provided him and successor kings with license to "visit, repress, redress, record, order, correct, restrain, and amend any errors, heresies, abuses, offences, contempts and enormities" in England's religious affairs. Such suffocating state control of religion wasn't what the Founding Fathers had in mind. They wanted a constitution in which religion would be free of government interference, while religion in turn would not attempt to impose its power on government—a clear separation of church and state. For the Founding Fathers, American governance was a secular experiment with its roots in the Enlightenment and the French Revolution. Government and religion were to be two separate magisteria. It was best to keep religion apart, since it was enough of a chore to write a constitution to prevent a democratic government from morphing into a secular autocracy.

In colonial times ceremonial public religious expression was commonplace and not a significant concern. Until World War II, Bible readings and recitation of the Lord's Prayer were the everyday practice in public schools. To most people it seemed that such moments of Christian tradition were a harmless way to help mold student values. Jehovah's Witnesses disagreed. Because they believed strictly in the injunction from the first of the Ten Commandments that "ye shall take no other Gods before me," Jehovah's Witnesses refused to salute the U.S. flag. This nonconformity led to persistent harassment and suspensions of their children from school. The Jehovah's Witnesses pursued their religious freedom through the courts but lost the Gobitis decision when the Supreme Count affirmed the statutory requirement to salute the flag. But only two years later in 1942 with three new justices on the Supreme Count, the Jehovah's Witnesses won a victory for religious freedom of expression carried by the vote of liberal (sic) judges (West Virginia v. Barnette).

In the realm of the separation of church and state, a series of Supreme Court decisions soon after World War II banned prayer and Bible reading in public schools. In more consequential matters, recent court decisions have favored government ladling out of tax money to religious organizations. In 2002 the Supreme Court ruled that government support of school vouchers was acceptable (Zelman v. Simmon-Harris 536 US639). At present about 95% of such tax money goes to religious schools. The federal government also gives millions of dollars to faith-based initiatives, which aid many denominations in their evangelical mission while requiring almost no accountability for spending. Flush with tax money, conservative religious organizations have resources to provide financial support for election campaigns though issue advertisements, the financing of voter turnout campaigns, and endorsements that encourage church members to donate to particular candidates. Catholic activist judges on the Supreme Court appear eager to breech Jefferson's wall of separation between church and state.

The fantasy of some conservative Christians is to restore the Christian heritage of the United States, which they claim has been stolen by revisionist liberals. Certainly, as a strategy, it would be much easier to restore a Christian theocracy than to attempt to start one from scratch. However, this sweeping notion of "taking back America" is fraudulent because, as we have seen, American governance was never Christian (Nettelhorst, 2006). The Founding Fathers were not untutored in European history. They were fully aware of the abuses in Europe's model of state-supported religion, predictions borne out as we have described, when the Nazi's exploited the financial dependence of churches to hobble Christian opposition. And the Founding Fathers knew that European Christian autocracies would surely seek to quash anything resembling the radical proposal of American governance centered on personal freedom and participatory democracy. These considerations made the Founding Fathers especially keen to exclude the use of public tax money in the support of religious enterprise.

The Christian Right is now maneuvering to gain its own vaunted theocracy—spurning compromise or accommodation with more moderate values in order that their particular vision of Christ shall rule. Representative Christopher Shays of Connecticut has said, "The Republican Party of Lincoln has become a party of theocracy (New York Times, March 23, 2005)." This would please the blunt-spoken Pentecostal Pat Robertson who said "there is no way that government can operate successfully unless led by Godly men and women under the laws of Jacob (quoted in Phillips, 2006)." In a theocracy one can expect to have science, medicine, education, social policy, and even wars guided by faith rather than by evidence, as has been the case during the presidency of George W. Bush. Effective democracy depends upon diverse views and thoughtful assessment of issues in contrast to faith-based politics, which employs absolutism, authoritarianism, and divine guidance. As President Bush said, "I trust God speaks through me. Without that I couldn't do my job. (Lancaster New Era newspaper, July 16, 2004)." Tom Delay, former House Majority Leader, now under indictment for voter fraud was even more emphatic. "God is using me all the time, everywhere, to stand up for Biblical world views in everything I do and everywhere I am. He is training me."

Historical records do not support the popular, yet incorrect view, that America was founded as a Judeo-Christian nation. As revealed in their own writings, many of the Founding Fathers of the United States were deists. Judicial rulings favoring school vouchers and millions of tax dollars in support of religious activities are a troublesome reactionary change that fuels the fires of theocracy in America. It will require a perceptive populace to understand the multiple risks of linking religion to tax dollars.

19

&

Onward

Living by Evidence

This has been a large undertaking—to see what a biologist who studies the brain can usefully relate about the nature of life on Earth. In the process I have simplified the religious landscape and underestimated the sophistication of spiritual individuals. Much of the biblical history and ageless theological arguments are familiar to the religiously informed. Even so, it can be painful to acknowledge the excesses and abuses racked up in religious history. Perhaps I have been too forthright in probing the origins of biblical scripture or raising questions about the probable character of Heaven with God as its overseer and the arbiter of prayers. And yet, to leave such problems unaddressed builds a weak foundation for the challenges of spiritual life.

The inaccessibility of the supernatural realm is the foremost constraint on religious beliefs. Claims about the supernatural rely largely upon scripture, revelation, and theological pronouncements that nourish the desires of individuals and fulfill the programmatic aims of religious institutions. A vast theological literature claims to know God's plan and actions, although what actually goes on in the supernatural realm resists verification. This is evident in the enormous diversity of religious denominations and in the failure of centuries of effort to reconcile the goodness of God with the harsh realities of life on Earth. Consider the following real-life challenge.

A Southern Baptist campus recently had an opportunity to

deal firsthand with the classic problem of evil and harm. On February 6, 2008 in Jackson, Tennessee, a fusillade of tornados raked Union University, or what used to be Union University. It took only a few minutes to flatten 32 of 33 buildings and destroy some 1,000 cars. The Governor said it was as if "God had used a Brillo Pad." Richard Fausset of the Los Angeles Times interviewed several grieving students and staff. They didn't understand why God let this happen, but considered it test of their faith. They thought it a sign of God's compassion that no one was killed outright. At another time and place equally religious Jews in ancient Israel would have concluded that vengeful Yahweh had become incensed about sinful violations of his commandments. Whether seen as God's test of faith, or God's anger at sinning, such precast rationalizations for God's apparent shortcomings are part of the foundation that stabilizes fundamentalist communities. Well-coached youths are resolute in the willingness to forgive God. But for many thoughtful people such clerical reassurances, shrugs of puzzlement really, can be too feeble to sustain an embrace with a God who allows such harm and evil (Barker, 1992; Ehrman, 2008).

Advances in science and technology create theological indigestion and erupt into gritty skirmishes over matters like the theory of evolution and beginning-of-life and end-of-life decisions.

Religious extremists and secular doubters have diametrically opposed views of evidence for divinity. For entrenched believers no evidence is necessary; for confirmed skeptics no evidence has been sufficient. Ecclesiastical declarations about supernatural features typically only succeed in raising more questions and additional "answers" in familiar spirals of speculation. This book has sailed far enough toward otherworldly oceans—it will not be drawn into whirlpools of theological futility. Those who want to embrace creedal assertions about the supernatural must either have outright faith or submit to the authority of scripture or priests—universally persuasive evidence is not at hand. These things said, there are notable instances when religious devotion has calmed life's troubled waters by providing comfort and sustenance.

Evolutionary Passages

Look at the intricate beauty of flowers, butterflies, and birds. Their apparent designs reflect the most popular argument for God's existence. Yet, no proposal of design by the divine can account for the complex evolution of biological structure and function. Quite unlike a highly Intelligent Designer, evolution has followed remarkably untidy paths in arriving at splendid adaptations of plants and animals to their environments. Fundamentalists must face a bracing conclusion: Intelligent Design repeatedly fails to account for biology's untidy realities. In contrast, evolution by natural selection succeeds as a proven explanation for life on Earth.

To restate briefly biological evolution's fundamental process, occasional variation in heritable genes modifies the offspring. Natural selection tests these slightly modified descendants for their fitness to their environmental circumstances. The better-adapted individuals are more likely to survive. For cultural evolution, which relies on passing along oral and written traditions to the next generation, language is the key. As cultural carryovers from superstitious and uninformed times, many early sagas and myths are flatly inaccurate. The much-improved understanding of the way evolution proceeds and organisms function provides modern humans with a clearer picture of life's processes. Consequently, many Christians now accept the facts and mechanisms of geological and biological evolution. However, others abhor the theory of biological evolution, as if it were a nasty bramble blocking their way toward Salvation and God's embrace. Fortunately, the devout need not cast off evolution in order to hold on to God—there is no logical difficulty with believing both in an active God and in evolution (Ruse, 2001; Judge Jones III, in the Kitzmiller v. Dover decision, 2005).

Let Humility Illuminate Man and God

The humble evolutionary history of humans. There is no easy way to shield the emergence of humans from the rough and tumble of evolutionary processes. If humans are God's chosen climax, why has human existence on Earth been so brief compared with the billion-year history of unicellular organisms? If humans are so important to God, why did it take so much time for us to emerge and eventually become well enough informed that we might possibly control our own destinies in a respectable manner? It is a vastly overrated conceit to remark, "We are the pinnacle of life." This claim might prompt an extraterrestrial observer to ask, "Take a look at the widespread poverty, the malnutrition of millions, the catastrophe of sexually transmitted diseases including AIDS, the ceaseless religious and secular bloody wars, the prevalent crime and drug addiction, the pervasive moral degeneracy, and the relentless destruction of the environment, including much of its spectacular plant and animal life. Do you consider this a legacy suitable for crowning as the pinnacle of success?" We need to ponder our good fortune that we humans exist at all. It is time to apply a balm of humility to relieve the religious itchings of our demanding egos.

God. Humans stirred to spirituality by their feelings, may seek God—especially a God willing to address personal concerns. Dialogs with God can pleasure our vanities. Even so, our notions of God must be tempered by considerable humility as we consider the origins and status of humans on Earth. The theory of evolution has replaced many superstitious creation myths.

One proposal for eliminating conflict between God and scientific laws is to consider that God is responsible for all the laws of nature. "In a real sense the laws of nature are the operation of His governance." (New Catholic Encyclopedia, 2nd Ed. volume 5, p. 495, 2002. Thomson, Gale Publishers, Detroit, Michigan). So, each of us must face a crucial question, "How much does God currently do?" Is there an activist God who carries out identifiable interventions consistent with the features of life on Earth? Pope John Paul II conditioned his belief in evolution on divine guidance because his pivotal concern was not with the historical fact of evolution but with its mechanisms. Was it natural selection or the adroit hand of God that steered the history of biological evolution? Physics, chemistry, and biology seem sufficient to explain the natural world without recourse to supernatural. If God does intervene with controlling influences, this adds some major complications. Straightaway, God appears to be accountable for the existence of widespread human pain and suffering (Ehrman, 2008). In addition, God's frequent interventions require the miraculous transfer of divine energy from the supernatural to the natural realm. Moreover, if meaningful free will also exists, it narrows any influence God would have upon ongoing human evolution since in generation after generation the traits of children stem from our free choice of reproductive mates, not from God's choices. Undaunted by all of these concerns, many theologians respond that God is above analysis and behind everything—science ought to cease its bickering and submit to God's authority.

Accepting an Understanding of Nature

Science relishes transforming mysteries into understanding. Religion relishes continuing mysteries that defy understanding. Although mystery is the message for each, the intentions differ. For spiritual individuals mystery provides the hope of a sublime afterlife. It is one thing for needs to arouse feelings of the divine inaccessible to science. But it is quite another to dismiss accurate geological and biological rules governing Earth and its life. The temporary comfort such denials afford cannot be sustained—the truth will be found out—hypocrisy will be found out. It is understandable but unproductive to castigate scientific advances that dash visions of a fancied afterlife. Rather than lambasting cosmologists and biologists for having deflated excessively optimistic religious promises, mature adults should press preachers to cleanse their creeds of scientific misinformation. As a first helpful step, the theory of evolution provides a broad understanding of the natural world.

For a more theological outlook, let's turn to a noted expert on comparative religion. Huston Smith began his lifelong love of religion as a child of Christian missionaries in China. After examining all of the world's great religions he concludes that God is a creative spirit filling the universe—an impersonal God that doesn't interact with Earth. In his book, *Why Religion Matters* (2001), Huston Smith reveals his deep emotional suffering. He plaintively, even desperately, seeks an afterlife, but fears there is none. I am pained that he blames scientists for deflating many of the myths that once lifted his hopes and buoyed his spirit. We ought not to attack scientific messengers for impartial investigations even if the results cast doubt on traditional narratives.

Perhaps Huston Smith has too grand a view of the scope of scientific investigation. The theory of evolution can't even investigate, let alone rule on, divine intervention. Scientific investigations of evolution can only rely on material or natural explanations because the supernatural is purely and simply out of the reach of science. Scientists do not know what properties may lie beyond measurable features like matter and energy. Humans have imaginations that can roam far beyond the reality we can access.

The Purpose of Human Life

Every culture imbues life with purposes and makes recommendations for how to spend one's life. For Buddhists, life is an exercise in surmounting suffering. The ancient Greeks thrived on exercising their reason. Absorbed with devotion, the Abrahamic religions (Jewish, Christian and Muslim) often attempt to discover and carry out God's will. The purpose of many Christian lives is to model their daily lives after the compassion and glory of Jesus. Where do you find your purpose in life? What should you do with your life? Is your purpose to optimize the welfare of your offspring, spend your efforts on creative or productive ventures, work hard to gain wealth and power, or follow a path of personal indulgence, possibly working only as needed. Or should you devote your energies to getting into Heaven, perhaps in the process even ignoring Earth's secular vale of tears by retreating to a monastery for contemplation and prayer?

God's aim for Earth. Claims about God's purposes can be puzzling. Why is it our role to do God's will on Earth, since God could set things right himself rather than depending upon a multitude of bumbling and squabbling humans? Why would he have some unfathomable needs to satisfy; some hidden objectives requiring the participatory struggles of flawed humans? Are we part of a morality play to test whether humans armed with free will can overcome self-absorption and the deadly sins? Is it to see whether God's kick-start of biological evolution provided the momentum for the emergence of sustainable peace and tranquility at the end of a billion year parade of some increasingly smart creatures? Saddled by the gift of contemplation, we either grapple with the complexities of our fate, or passively accept those theological answers we learn by rote.

Consider that God's plan for life on Earth may be an open future that allows for new and varied forms of life whose character is not predetermined. If so, God's plan may be fulfilled in many ways. God may be satisfied with various future outcomes on Earth, few of which would involve humans as permanent inhabitants. Late on the scene, we are likely to be a passing phase. If we follow the history of so many other species, we will become extinct in a few million years at best. If we continue with our present numbing despoliation of the environment, it will be ruinous for humans much sooner. All of this could occur if God's plan depends on heritable genetic variation selected by fitness; that is, if God's plan is the sort of evolution that has been going on for more than three billion years.

Secular views of Earth. The outlook of biological scientists is to marvel at the intricate order of nature and to appreciate the vast chemical networks that synthesize molecules and give cells their self-regulating vitality. Scientific understanding replaces the fog of mystery with a myriad of bright, purposeful, biochemical reactions. It gives enormous meaning to understand the intimate mechanisms of biological life—to discover its functional rules. Scientists are answering so many questions every day that they sense a giant illuminated book of living networks opening before them. The orderly beauty of life and its resilience are realities that sustain the spirituality and reverence of scientists. Somewhere deep in the vastness of space there may be grander lives and evidence of greater sophistication, but life on Earth is sublime enough for reverence, even adoration. Somewhere too there may be stupendous supernatural forces whose properties we cannot examine and therefore cannot fathom. These forces may be of unimaginable complexity. However, with deepening understanding, it has become evident that many organisms on earth lead lives sufficiently beautiful for both scientists and non-scientists to appreciate.

The Meaning of Human Life

Sustaining the Earth's environment. Many denominations—Muslim, Orthodox Judaism, Catholicism, Hutterite, Mennonite Protestants, and others—are vigorous advocates of high rates of reproduction in order to swell the ranks of their denominations. Biologically, enthusiasm for reproduction is hardly surprising. Unfortunately, by our abundance we humans are stripping the planet of natural resources at prodigious rates. So is the real meaning of temporal life to gain access to the hereafter rather than worrying about sustaining the environment's remaining resources? Life on Earth is clearly at risk if the lure of Salvation convinces the powerful few or the plentiful multitudes that our temporary stay on earth is an inconsequential inconvenience because the truly devout will escape in a Heavenly afterlife. Let's be clear: indifference to the environment today means indifference to the welfare of your own descendants as their future unfolds with ever increasing struggles.

Of course, most people believe that every human should live well—or at least that no one should be impoverished. Those economists, who worship the mantra of growth, see no insurmountable barriers to continued development and increased consumption, which they imagine will result in a marked rise in the standard-of-living for the struggling 80% of the world population. Biologists believe these rising demands of the expanding human population will exceed the carrying capacity of the earth. The wretched multitudes deserve a better fate than they are likely to get. The signs of strain are all around us. Touting optimistic eco-fixes and techno-fixes, politicians are unwilling to address the issues of unsustainable environmental exploitation driven by the looming desires and aspirations of billions of consumers-to-be when they cheerfully unleash a ten-fold increase in the worldwide demand for natural resources. Captive to our primal urges to multiply and exploit, we relentlessly push the envelope of risk and profit. There are many grand intentions but few realistic plans to halt the relentless and unsustainable over-exploitation of the environment by an enormous population (Martin, 2006).

The meaningfulness of life on Earth. It might provide some structured meaning, if we could identify a biological plan that controls our lives. It turns out that evolved human life operates with a remarkably detailed and resilient plan and you are an essential player. It features specific error-correction mechanisms to ensure great precision. Our lives begin with one of the most remarkable feats in the known universe—the embryonic development of a human. As Rabbi Maimonides (1135-1204) appreciated, it would be a miraculous process if an angel guided human development from egg to child. Even so, in his other role as a physician, he realized it would be even more impressive if the embryo were capable of guiding its own development without divine intervention. Consequently, for free, with no input beyond sensible nutrition, the embryo's developmental plan builds you a body with a splendidly creative brain.

I challenge those who sniff incredulously at humanism to spell out a deeper and more profound purpose than individual human creativity and productivity aimed at righting the world's wrongs. The everyday life of many fortunate humans has a specific purpose which is to be creatively productive, as best we can arrange it. If wisdom prevails, we can expect human creativity to spread material comforts and good health. Creativity satisfies deep needs; humans have been creating durable products like artwork long before writing emerged to provide a historical record of human activities. Whether to attempt to burst on the scene with a dramatic flare or to make smaller, more sustained, contributions—it's your choice. It has been said with some accuracy that over the centuries, music, language and poetry, and some forms of art will outlast buildings, governments, and even nations. But if durability is the hallmark of worth, then the longevity of scientific discoveries and inventions are among the most permanent creations of humans. The most enduring creations are novel contributions to understanding, because a discovered truth is virtually indestructible. To feel one has made a durable contribution, however esoteric, can provide a larger-than-life boost to the ego while quelling feelings of insignificance. Basic research makes unique discoveries that forever help to reveal the nature of life and the universe. Mathematicians and engineers create concepts and inventions that may be widely adopted. Naturally, there are many other paths to creative accomplishments. Literatur sic, and the arts provide arenas for durable and unique crea The health professions, including nursing, public health, an technology make direct and proud contributions to reproductive and child health. These increase the odds that our genes will be successfully passed on to the next, healthier generation—a clear contribution to humanity.

Some assert that human evolution fails to offer an ultimate meaning to our lives. Life must lack meaning if it has simply evolved—or perhaps not so simply—with no purpose beyond persistence. Faced with this somewhat bleak assessment, it is natural enough for inquiring minds to examine the diverse meanings offered by religion. But to make a proper examination we must shed the rosy glasses of faith, open our eyes, and be prepared to ask questions. When alternative religious meanings of life are offered, we must ask which one of those meanings is actually correct. Do we stop searching when we encounter the meaning we like—perhaps an easy path to the afterlife?

Salvation. The most desired feature of Christian Salvation is release from the clutches of earthly depravity to receive a Heavenly afterlife. In this regard, evangelical branches of Christianity have had a mission to save as many souls as possible. This can be a risky enterprise. Surveys generously estimate that in 2000 CE alone nearly 100,000 Catholics and 60,000 Protestants were killed in the line of duty, all martyrs to their faith (Barrett et al., 2001). Over the centuries there have been millions of martyred soldiers of the cross, dead in trying to subdue disbelievers, who died in much greater numbers. As we note the courage and dedication of Evangelists, we should not overlook the consequences of their determination to turn out religious converts. The fallout includes support of slavery and servitude, the willful extermination of native cultures, religious wars, and other harsh impositions including imprisonment and ensnaring the vulnerable young—all weapons in the Christian arsenal to save souls.

The meaningfulness of life in Heaven. It would seem that any religion whose prime objective is entry into Heaven takes on a substantial burden both of justifying its activities on Earth and explaining how everlasting life provides a self-evident purpose. Is it the purpose of human life on Earth and the end-point of evangelism merely to ensure everlasting leisure in Heaven? For some of us such an aim has no more substance than a daily selection of a few diversionary DVDs. What is profound about an afterlife that resembles a permanent vacation where one lounges poolside playing cards, listening to music, and sipping cool drinks for an eternity? If we hang out in Heaven, do we want to play Bingo forever? Many have drawn great satisfaction from work either as a creative outlet or in its capacity to help others. After working productively for decades followed by a period of rather aimless retirement before death, you might like to work once again. What activities in Heaven are more meaningful and fulfilling than the earlier charitable opportunities to help unfortunate individuals who populate the earth? Not that anyone claims Heaven ought to be supplied with destitute individuals so the rest have someone to look after. Or would it be your intention to spend an eternity lounging in Heaven with no responsibilities or challenges to get up for in the morning? How many lattes can one drink in an eternity, or how many millions of rounds of golf can one play and replay? Self-indulgence gets old. Perhaps everyone "does their own thing" in Heaven. If so, then eternity's meaning must be whatever you wish to make of it—lounging or laboring according to personal whim. Those smitten with their Salvation need to spell out exactly what they are saved for and how they know it.

Strategic hedging. Skeptics often consider hedging their bets by planning as if Heaven really exists. Why risk missing out? Alternatively, perhaps your needs, hopes, fears, and reasoning have already led you to believe you are headed for Heaven. There is comfort in that. But in that event, hedge your bets in the other direction and continue to direct the utmost respect and effort toward lives on Earth—it may just be the only life you and your kin will ever have.

Immortality Revisited

Science has nothing to say about eternal life. However, it can predict that in a few million years, humans as presently constructed, will not be the dominant form of life on Earth, and in five billion years no earthlings will be bound for Heaven because our expanding sun will have vaporized our splendid Earth into gas. However, since science may prolong our lives, let's get our bearings by returning to earth as it is now.

As is evident from three different perspectives, earthlings are continuing their search for guaranteed bodily immortality. We have already considered immortality from the Christian perspective. According to Genesis, it was God's intention for Adam and Eve to be immortal, but they scotched this plan by romping around and misbehaving in the Garden of Eden. Thereafter, no human has been born immortal. With the coming of Jesus, human souls and bodies received renewed hope for immortality.

In another tilt toward reduced mortality, medicine is presently extending human lives. Suppose that new medicines give humans several more decades of life, so that apart from a mishap we are somewhat less mortal than before. Of course, as a centenarian you also want to be spry, for who wants to be a relentlessly withering cripple? The elderly hope that medical advances will at least permit them to retain all of their teeth and to continue to hear quiet conversations.

A third perspective on immortality is the paradoxical gratitude we owe to death. We and other large multicellular creatures that reproduce sexually have well-defined lifespans. Short lives can be helpful. Every act of sexual procreation creates a novel assortment of genes with its slim chance of being better than its parents in making a living in a particular niche. In generation after generation, over hundreds of thousands of years of births and deaths, natural selection tests different assortments of genes seasoned with occasional mutations. This repeated testing has resulted in highly sophisticated animals and plants, marvelously adapted to their environments. Imagine how evolution might have stalled-out if multicellular organisms had become immortal and essentially ceased reproducing. It is impossible to express sufficient gratitude to the long line of dead ancestors whose adept environmental adaptations and useful mutations permitted some really neat genes to be passed on to us.

Life's Transcendent Narrative

Nevertheless, it is clear that human spiritual needs won't be satisfied solely by honoring the debts of gratitude we owe to the cunning and mortality of our many remote ancestors. Humans also need panoramic narratives—heroically uplifting narratives that not only trace our past but also give ample hope for the future.

Without exception, the saga of organic evolution is the universal narrative of life's past. Woven with a few basic rules, evolution's narrative tapestry depicts remarkable acts of dedication and heroism rich in beauty and detail. Whether it is the protective coloration of fawns, the trilling of a wood thrush, the twinkling of fireflies, or the savory scent of an orange—life's features have splendid purposes. Life is flooded with purpose. Organisms are interlinked collections of functions—purposeful muscles, receptors for colored light, for sounds, and for smells—even a single molecule may attract a mate. You say you have been looking for purpose in life. Well, what else is all around you if not organisms with a vast menu of purposes? The phrase, "my life has no purpose" is an empty cliché. In the short run the clear purpose of our lives, following our genetic mandate, is to make what we can of our opportunities, whatever they are. Develop your potential. Oh, you want to go back, far back, to some ultimate purpose for primates, or for mammals, or for amphibians, or for single cells, or for life or...? Well, remember when you delve into early life and then go still further back, you will reach a point where neither science nor religion can tell you why there was something there, rather than nothing at all.

Immortal Genes?

Imagine you have spent your own life well. What kind of a future will you have beyond that? In the intermediate term a further purpose knitting up the lives of most humans is the extraordinary opportunity to help our genes. In searching for personal immortality for your shopworn body, you may lose sight of the partial immortality of your genes themselves. First, let's give genes more credit—much more. They are responsible for the construction of your body, including the circuits of your brain that contribute to your remarkable ability to sense color and musical harmony, or to feel a tender touch, and in a different way, to feel joy.

In a small step toward the immortal, a normal mating passes on to each offspring about half of each parents' genes. It is a splendid blend that is not, however, a random alphabet soup of genes. Genes are passed on in clusters. This is one reason why many of our traits have a visibly extended lifespan of several generations. And our individual genes are likely to be around for many thousands of years, especially the genes crucial for human reproduction and survival. Our genes should thank us for our wise parenting. In 10,000 years the family of human genes will still be 99% like they are today. So in spite of some mutations and disruptive chromosomal rearrangements, your genetic information is quite stable. It is simply tumbled about and dispersed across your line of descendants. Your children will inherit roughly 50% of your genes, your grandchildren 25%, and your great-grandchildren about 12.5%. The initial clear resemblance to yourself may diminish with successive generations. This is probably for the better. Few of us have such a desirable combination of genes that it wouldn't benefit by infusion of fresh supplies—a little hybrid vigor. Nonetheless, it is pleasing to imagine that everyone in a very long line of your descendants will share a part of you, without exception.

As noted earlier, your quasi-immortal genes use your body as a temporary safe harbor, a refuge before the genes move on to a newborn child who shelters them in the next generation. Some individuals may feel rather callously used by their genes, even if in our fleeting lives we take great satisfaction in the child who is their next repository. Does it annoy you that your genes, rather than you, have some measure of immortality? "What about me, don't I matter?" cries your ego. But it is the genes' tale to tell. Individuals are temporary products of life's relentless forward flow of genes.

A Path Toward Faith?

In reviewing this book's journey we found that the Bible was assembled in order to further the objectives of a church whose leaders strategically quashed abundant alternative views and writings. Nonetheless, traditional Bibles contain useful historical accounts and lessons of life that warrant close reading. Still, in our spiritual search we need to remind ourselves that our most secure roots lie in our indelible biological heritage, not in versions of ancient myths crafted to billow the doctrinal sails of an institutional church.

When examining ancient customs we encounter some treatments now considered onerous and cruel. It is fair to say that the history that I recounted here sometimes emphasizes repellant actions and harsh injunctions prompted by religious beliefs. These history lessons might be unnecessary had modern humans been able to rise above such mean and sanctimonious postures of superiority. Sadly, it is still true today that, "Men never do evil so completely and cheerfully as when they do it from religious conviction" (quotation attributed to the devout Christian, Blaise Pascal [1623-1662]).

It is not possible to rule out the existence of a supernatural God. For myself, I am just as keen as the next person to get to Heaven under the protection of a benevolent God. As a neuroscientist I am inclined to believe that if there is a Heaven, it is most likely populated by our immaterial spirits. That said, as a matter of practicality I assume my present life on Earth is the only one I will ever have. I am trying to make the very best of it for all concerned. I hope you will see this book as my effort to stress that evidence is the way to illuminate the paths of your own journey—glistening facts free of moldy dogma.

My mission in this book has been neither to dismiss responsible religion nor to offer only science, but to try to give an honest account that addresses existential issues. I am not interested in presenting a dogmatic vision of the correct path or the correct value, except to say that demonstrable truth, not assertions of truth, matters deeply. The best means we have for factual discovery is empirical science—its results provide precision and predictability. You deny yourself wisdom if you pursue life with indifference to an evidence-based education, and instead accept a fanciful account from the innumerable merchants of faith. Few attitudes present more danger to society than acting on speculations as if they were true. Faith can weaken communities by estranging people from education. Arrogant claims of doing God's will can lead to self-serving behavior and a litany of evil actions that could not otherwise be justified. Faith can bring personal fulfillment but at the cost of social irresponsibility. As the great historian Will Durant (1950) observed: "Intolerance is the natural concomitant of strong faith; tolerance grows only when faith loses certainty; certainty is murderous."

Evidence-based truth is the only proven path—all else is vain hope. While evidence should trump self-indulgent needs, it is important to consider experimental evidence as yielding provisional rather than absolute truth. Much of the function of science is to make provisional understanding more accurate. Since all of us must admit we live in personal ignorance of most of the things already well-understood by specialists, surely we can live with the uncertainties about life that have not yet been clarified by scientific observation. Absolutism is a dreadful condition; it is the worst feature of faith. The absolutism of faith is only a display of the intensity of personal needs—it is not in any sense a display of information about reality. Shouting something is true, even singing it up to the Heavens, adds intensity but not credibility. The attempt of the faithful to create truth by strenuous exertions is an impossible dream—an exercise in wish fulfillment. Faith-based convictions offer a comforting cocoon that is but a hollow refuge. Unlike the reassuring supernatural beliefs offered by rigid faith, evidence-based truth may shift prior beliefs. It is part of the discovery and maturation process to rebalance one's stance to a more certain understanding. As Martin Luther said, "Everyone must do his own believing as he will have to do his own dying."

Spirituality on This Pale Blue Dot Called Earth

Had physicists relied solely on basic principles of energy and matter, they would not have predicted the existence of complex life forms. The existence of any self-sustaining living system is a thermodynamic surprise. As a defining mantra of physics, the law of thermodynamics says that over time the entropy of matt increases, as represented by greater disorder. Yet, living systems, representing rare pockets of ordered molecular systems, can exploit available sources of energy to become even more orderly. Clearly, it is possible for matter on a small scale, notably living organisms, to become increasingly orderly by exporting disorder. It is a surprising transcendent reality that energy can empower the creation of order. The complex forms of life's endless beauty are both understandable and precious to us. Self-organization is a core principle of all life. Our spirituality, our hope, lies in self-ordering systems that are the essence of life. Our brain is at the core of the matter. Neuroscientists, like myself, marvel at the astounding capacities of even the few features of the brain we presently understand. In smelling a single odor molecule or seeing one quantum of light, brains are capable of exquisite sensory discriminations that exceed the sensitivity of physical and chemical assays. We can learn associations and understand subtle relationships that baffle computers. Children acquire information and complex language with stunning ease and rapidity. Most providentially, the more we use our brain the better it performs—practice elevates skills. The monumental challenge is to put our brains to good use. That will require deep wisdom and more control over our profligate emotions than humans have demonstrated in the past.

Subtle principles at work in the most particular of physical circumstances have allowed life to seed itself on our planet. Here on Earth exemplary people offer hope that Earth can become an oasis of civility—a divine speck in a cosmological tableau peppered with hot stars but otherwise vast, dark, and frozen almost to atomic immobility. Saving Earth is the objective of many Christians so that through concerted efforts Earth will become a Heaven. That we are able and eager to embrace this prospect at all is evidence for spirituality and religiousness. The prime intention of all sacred quests is to under gird our lives with a sustaining spiritual world. Even without outside supernatural assistance, on-going and anticipated understanding offer some hope that Heaven will come to reside on Earth. The 20th century Protestant theologian, Paul Tillich, concluded that God is not a supreme being out there in the great beyond. God is the Essence of life that gives us the capacity to forgive, reconcile, and transform for the better. It is most vexing to admit the odds are long that good shall triumph over human self-concern and the prevalence of evil. It is one of the great paradoxes of life that in spite of natural selection for long-term survival, the behavior of humans and their governments caters nearly exclusively to short-term satisfactions. Moreover, we must confront swaggering reactionary forces that both display and exploit profound human ignorance and superstition.

What we need most from the spiritual is hope for the future. Hope that we will use our manifest talents for the betterment of life on Earth. Hope is a primal necessity. Through meditation and prayer one can feel and embrace the aura of goodness that pervades each person. The desire and dignity in each person can be strengthened to do that which is right. When physicists say that everything around us, everything we sense, is some form of energy, it reminds us that there is energy enough to transform Earth into Heaven. We will need energy to diminish hatreds and embrace love.

It is consummate hubris for any human to create, and there is no other word for it, to create a supernatural God who encompasses all space and time, all past, present and future knowledge, a limitless God of immaculate perfection. The intention is clear, "and I so revel in the importance of my needs and feelings that I must have a personal God who can never in any way be limiting for my situation." To assert further that this is the God all others must accept is a profound moral and intellectual failure. A moral failure because it is so extraordinarily self-elevating—no ordinary individual, like myself, warrants an entire universe of special consideration. An intellectual failure because standing on the thinnest of evidence the proposal obliterates what we know and have learned about the natural world. It obliterates all the products of education and civilization for the sake of human ego. It hurls life on Earth not into light of clear understanding but into the darkness of rank conceit. Whether religious or not, reasonable people will conclude—indeed must conclude—that a life based upon empirical fact is not compatible with a life dominated by faith. Is there any more momentous choice for one's family? At the end of the day, at the end of a life, which of God's children deserved being reared in the darkness of deception?

In the natural world scientists are presently engaged in detailed examinations of genes in the context of evolutionary change. Factual accounts are replacing mythical accounts. Scientific truth is in ascendancy. The long shadows of Dark-Age mysteries and speculative misconceptions are in retreat. Tattered medieval parchments displaying bizarre bestiaries in a fanciful Hell run by a mythical Satan are properly regarded as trumped up fears. Satan has always lurked in the dank recesses of suffering and misery. Now at last in our generation the glow of scientific discoveries has illuminated and purified most of Satan's dark lairs with the light of understanding. Gone are his traditional haunts in the mentally possessed—persons now known to be afflicted with epilepsy, palsy, or psychosis. As shrouds of ignorance dissipate, reclusive Satan retreats. No longer does he squeak and gibber in dark corners. Only a few hiding places remain—it is time to say Bye Bye Satan, forever.

When most of the stars have lost their blue-white sparkle and faded into a dull red twilight, the universe may be more than a trillion years old. Even so, somewhere it will be inhabited by evolved creatures of inconceivable sophistication, perhaps nearly as powerful as the Gods early humans so hopefully imagined in the late Earth's own time. As these advanced beings sense their end is also drawing near, they may nostalgically yearn for earlier and simpler times—times poignantly much like this exquisite instant of yours and mine, when our generation decoded the secrets of our genes. Our personal days are just a fleeting instant in deep time—but for those minds seeking spiritual understanding, right now is the most exciting moment in Earth's five billion year history because now at last we are coming to understand life on Earth.

References

Ahlberg, P. E. and J. A. Clack, 2006, A firm step from water to land, Nature 440: 747-749.

Alaya, Francisco, 1998, "Human Nature One Evolutionist's View" an essay In: Whatever Happened to the Soul?, Walter. S. Brown, Nancy Murphy, and H. Newton Malony, Eds., Fortress Press, Minneapolis, p. 31

Armstrong, Karen, 1993, A History of God, Ballantine Books, Random House, New York, 460 pp.

Armstrong, Karen, 2000, The Battle for God: A History of Fundamentalism, Ballantine Books, Random House, New York, 442 pp.

Aristotle, History of Animals, Book VII, translated 1883, London, G. Bell & Sons, p. 583.

Azari N. P., J. Nickel, G. Wunderlich, M. Niedeggen, H. Hefter, L. Tellman, H. Herzog, P. Stoerig, D. Birnbacher, and R. J. Seitz, 2001, Neural correlates of religious experience, European Journal of Neuroscience 13: 1649-1652.

Barbour, Ian G., 1997, Religion and Science, Harper, San Francisco 368 pp.

Barbour, Ian G., 2000, When Science Meets Religion, HarperCollins, New York, 205 pp.

Barker, Dan, 1992, Losing Faith in Faith, Freedom from Religion Foundation, Madison, Wisconsin, 392 pp.

Barrett, D. G., G. T. Kurian, and T. M. Johnson, 2001, World Christian Encyclopedia, 2nd edition, Vol. 1, Oxford University Press, New York, 1730 pp.

Baigent, Michael, 2006, The Jesus Papers, HarperSanFrancisco, 321 pp.

Baigent, Michael and Richard Leigh, 2000, The Inquisition, Penguin Books, London, 336 pp.

Benson, H., J. A. Dusek, J. B. Sherwood, P. Lam, C. F. Bethea, W. Carpenter, S. Levitsky, P. C. Hill, D. W. Clem, Jr, M. K. Jain, D. Drumel, S. L. Kopecky, P. S. Mueller, D. Marek, S. Rollins, and P. L. Hibberd, 2006, Study of the therapeutic effects of intercessory prayer in cardiac bypass patients: A multicenter randomized trial of uncertainty and certainty of receiving intercessory prayer, America Heart Journal 151: 934-942.

Bentall, R. P., 1990, The illusion of reality: a review and integration of psychological research on hallucinations, Psychological Bulletin, 107, 82-95.

Bethge, E., Ed., 1972, Dietrich Bonhoeffer: Letters and Papers from Prison, p. 311, Macmillan Press, New York, 437 pp.

Bevans, C. I., Ed., Treaties and Other International Agreements of the United States of America 1776-1949, Vol. 11: Philippines-United Arab Republic, Washington D.C., Department of State Publications, 1974, p. 1072.

Bininda-Emonds, O., M. Cardillo, K. E. Jones, R. D. MacPhee, R. M. Beck, R. Grenyer, S. A. Price, R. A. Vos, J. L. Gittleman, and A. Purvis, 2007, The delayed rise of present-day mammals, Nature 446: 507-512.

Birney, E. and the ENCODE Project Consortium, 2007, Identification and analysis of functional elements in 1% of the human genome by the ENCODE pilot project, Nature 447: 799-816.

Bivens, A. J., R. A. Neimeyer, T. M. Kirchberg, and M. K. Moore, 1994-5, Death concern and religious beliefs among gays and bisexuals of variable proximity to AIDS, Omega: The Journal of Death and Dying, 30: 105-130.

Blanke, O. and S. Arzy, 2005, The out-of-body experience: disturbed self-processing at the temporo-parietal junction, Neuroscientist, 11: 16-24.

Bodmer, R., 1993, The gene *tinman* is required for specification of the heart and visceral muscles in *Drosophila*, Development, 118: 719-729.

Borg, M., 1994, Profiles in scholarly courage: early days of New Testament criticism, Bible Review 10: 40-45.

Bottoms, B. L., P. R. Shaver, G. S., Goodman, and J. Qin, 1995, In the name of God: A profile of religion-related child abuse, Journal of Social Issues, 51: 85-111.

Bowen, Catherine Drinker, 1966, Miracle at Philadelphia: The Story of the Constitution Convention, Book-of-the-Month-Club, New York, pp. 125-126.

Bowie, Fiona, 2006, The Anthropology of Religion: An Introduction, Blackwell Press, Malden, Massachusetts., 332 pp.

Bowles, S., 2006, Group competition, reproductive leveling, and the evolution of human altruism, Science 314: 1569-1572.

Brosnan, S. F. and F. B. M. de Waal, 2003, Monkeys reject unequal pay, Nature 425: 297-299.

Brown, D.E., 1991, Human Universals, Temple University Press, Philadelphia, 220 pp.

Brown, Scott G., 2005, Mark's Other Gospel, Wilfrid Laurier University Press, 384 pp.

Buchan, J. R. and R. Parker, 2007, The two faces of miRNA, Science, 318: 1877-1878.

Campbell, Jeremy, 2006, The Many Faces of God, W. W. Norton, New York, 314 pp.

Carroll, Sean B., 2005, Endless Forms Most Beautiful, W.W. Norton, New York, 350 pp.

Chadwick, H., 1991, translator of Confessions of St Augustine, VIII vii (17), p. 145, Oxford University Press.

Chapman, Bruce, 1998, Postscript In: Mere Creation, W. Dembski, InterVarsity Press, Downer's Grove, Illinois, 475 pp.

Chen, I. A., 2006, The emergence of cells during the origins of life, Science 314: 1558-1559.

Coates, M. I. and J. A. Clack, 1991, Fish-like gills and breathing in the earliest known tetrapod, Nature 352: 234-236.

Collins, Francis S., 2006, The Language of God: A Scientist Presents Evidence for Belief, Free Press, New York, 294 pp.

Cowen, D. A., 2004, The upper temperature of life--where do we draw the line? Trends in Microbiology, 12: 58-60.

Cox, J. J., F. Reimann, A. K. Nicholas, G. Thorton, E. Roberts, K. Sprinell, G. Karbani, H. Jafri, J. Mannan, Y. Raashid, L. Al-Gazali, H. Hamamy, E. M. Valente, S. Gorman, R. Williams, D. P. McHale, J. N. Wood, F. M. Gribble, and C. G. Woods, 2006, An SCN9A channelopathy causes congenital inability to experience pain, Nature 444: 894-898.

Darwin, Charles, 1959, The Origin of Species, reprinted in 2003, Penguin Group, New York, 496 pp.

Dawkins, Richard, 1986, The Blind Watchmaker, Penguin Books, London, 332 pp.

Dawkins, Richard, 2003, The Devil's Chaplain: Reflections on Hope, Lies, Science and Love, Houghton Mifflin, Boston, Massachusetts, 263 pp.

Dawkins, Richard, 2006, The God Delusion, Houghton Mifflin, Boston, Massachusetts 406 pp.

De Landa, Diego (1572) An Account of the Things of the Yucatan, translated by David Castledine in 2000, Monclem Ediciones, S.A. Mexico, D.F.

Doolittle, R. F., 1993, The evolution of vertebrate blood coagulation: a case of Yin and Yang, Thrombosis & Haemostasis, 70:

24-28.

Dunbar, Robin, 1995, The Trouble with Science, Harvard University Press, Cambridge, Massachusetts, 213 pp.

Durant, Will, 1950, The Age of Faith, reprinted by Easton Press in 1992, p. 784

Durkheim, E. 1915, The Elementary Forms of the Religions Life: A Study in Religious Sociology, translated by J. W. Swain, Allen and Unwin, London.

Ehrman, B. D., 2003, Response to Charles Hedrick's Stalemate, Journal of Early Christian Studies, 11: 155-163.

Erhman, Bart D., 2008, God's Problem: How The Bible Fails to Answer our Most Important Question—Why We Suffer, HarperOne, 304 pp.

Ehrsson, H. 2007, The experimental induction of out-of-body experiences, Science 317: 1048.

Falk, Darrel R., 2004, Coming to Peace with Science, InterVarsity Press, Downer's Grove, Illinois, 235 pp.

Flamm, B., 2006, Magnetic therapy, a billion-dollar boondoggle, Skeptical Inquirer, July/August, 30: 26-28.

Frazer, James G., 1915, The Golden Bough: A Study in Magic and Religion, 3rd edition, reprinted in 1990 by St Martin's Press, New York.

Futuyma, D. J., 2005, Evolution, Sinauer Associates, Sunderland, Massachusetts, 602 pp.

Gazzaniga, Michael, 2005, The Ethical Brain, Dana Press, New York, 201 pp.

Gehring, W. J., 2002, The genetic control of eye development and its implications for the evolution of the various eye-types, International Journal of Developmental Biology, 46: 65-73.

Gingerich, P. D., M. Haq, I. S. Zalmout, I. H. Khan, and M. S. Malkani, 2001, Origin of whales from early artiodactyls: hands and feet of eocene protocetidae from Pakistan. Science, 293: 2239-2242.

Goldberg, Michelle, 2007, Kingdom Coming, The Rise of Christian Nationalism, W. W. Norton, New York, 253 pp.

Goodenough, Ursula, 1998, Reverence for Life, Oxford University Press, New York, 197 pp.

Gould, Stephen J., 1999, Dorothy, It's really Oz, Time magazine, August 23rd.

Greeley, A., 1993, Religion and attitudes toward the environment, Journal for the Scientific Study of Religion 32: 19-28.

Green, R. E., J. Krause, S. E. Ptak, A. W. Briggs, M. T. Ronan, J. F. Simons, L. Du, M.

Egholm, H. M. Rothberg, M. Paunovc, and S. Pääbo, 2006, Analysis of one million base pairs of Neanderthal DNA, Nature 444: 330-336.

Flynn, Tom, 2007, The New Encyclopedia of Unbelief, Prometheus Books, Amherst, New York, 897 pp.

Gaidos, E., N. Haghighipour, E. Agol, D. Latham, S. Raymond, and J. Rayner, 2007, New worlds on the horizon: earth-sized planets close to other stars, Science 318: 210-213.

Green Ruth H., 1979, The Born Again Skeptic's Guide to the Bible, Freedom From Religion Foundation, Madison, WI. 439 pp.

Gross, Paul R. and Norman Levitt, 1997, Higher Superstition: The Academic Left and Its Quarrels with Science, Johns Hopkins University Press, Baltimore, Maryland, 348 pp.

Hamer, Dean, 2004, The God Gene: How Faith is Hardwired into our Genes, Doubleday, New York, 241 pp.

Hamilton, W. D., 1964, The genetical evolution of social behavior, Journal of Theoretical Biology, 7: 1-16.

Harris, Sam, 2004, The End of Faith: Religion, Terror, and the Future of Reason, W. W. Norton, New York, 336 pp.

Haught, John F., 2007, Christianity and Science: Toward a Theology of Nature, Orbis Books, Maryknoll, New York, 208 pp.

Hedrick, C. W., 2002, The thirty four gospels: diversity and division among the earliest Christians, Bible Review, 18.3: 20-31.

Hedrick, C. W., 2003, The Secret Gospel of Mark: Stalemate in the Academy, Journal of Early Christian Studies 11: 122-145.

Hitchens, Christopher, 2007, God is not great, Twelve, New York, 307 pp.

Hitler, Adolph, 1925, Mein Kampf, translated by R. Manheim, 1971, Houghton Mifflin, Boston, Massachusetts, 720 pp.

Hodge, Charles, 1874, What is Darwinism? Princeton University Press, Princeton, New Jersey, p. 142.

Hodgkinson, V. A., M. Weitzman, and A. D. Kirsch, 1988, From belief to commitment: The activities and finances of religious congregations in the United States: Findings of a national survey, Washington, D.C.

Huesmann, L. Rowell, 2007, The impact of electronic media violence: scientific theory and research, Journal of Adolescent Health, 41: S6-S13.

Jeffery, Peter, 2007, The Secret Gospel of Mark Unveiled: Imagined Rituals of Sex, Death, and Madness in a Biblical Forgery, Yale University Press, New Haven, Connecticut, 340 pp.

Johnson, Paul, 1976, A History of Christianity, Simon and Schuster, New York, 556 pp.

Johnson, Phillip E., 1993, Darwin on Trial, 2nd edition, InterVarsity Press, Downers Grove, Illinois, 220 pp.

Johnson, Phillip E., 2002, The Wedge of Truth: Splitting the Foundations of Naturalism, InterVarsity Press, Downer's Grove, Illinois, 192 pp.

Johnstone, Ronald L., 1997, Religion in Society: A Sociology of Religion, 5th edition, Prentice Hall, New York, 372 pp.

Jonas, Wayne B. and Jennifer Jacobs, 1996, Healing with Homeopathy: the Complete Guide, Warner Books, Clayton, Victoria, Australia, 347 pp.

Kirsch, Jonathan, 2006, A history of the end of the world: how the most controversial book in the Bible changed the course of Western civilization, HarperSanFrancisco, San Francisco, 340 pp.

Kirschner, Marc W. and John C. Gerhart, 2005, The Plausibility of Life, Yale University Press, New Haven, Connecticut, 314 pp.

Klopfer, F. J. and W.F. Price, 1979, Euthanasia acceptance and rejection related to belief and other attitudes, Omega, 9: 245-253.

Knauth, L. P., 2005, Temperature and salinity history of the Precambrian ocean: implications for the course of microbial evolution, Palaeogeography, Palaeoclimatology, Palaeoecology, 219: 53-69.

Kosfeld, M., M. Heinrichs, P. J., Zak, U., Fischbacher, and E. Fehr, 2005, Oxytocin increases trust in humans, Nature 435: 673-676.

Kramnick, Issac and R. Lawrence Moore, 1996, The Godless Constitution, W. W. Norton, New York. 208 pp.

Kreeft, Peter, 1993, Christianity for Modern Pagans, Pascal's Pensées Edited, Outlined and Explained, Ignatius Press, San Francisco, 341 pp.

Krucoff, M. W., S. W. Crater, D. Gallup, J. C. Blankenship, M. Cuffe, M. Guarneri, R. A. Krieger, V. R. Kshettry, K. Morris, M. Oz, A. Pichard, M. H. Sketch, Jr., H. G. Koenig, D. Mark, and K. L. Lee, 2005, Music, imagery, touch, and prayer as adjuncts to interventional cardiac care: the Monitoring and Actualisation of Noetic Trainings, Lancet 366: 211-217.

Lalueza-Fox, C., H. Roempler, D. Caramelli, C. Staubert, G. Catalano, D. Hughes, N. Rohlan, E. Pilli, L. Longo, S. Condemi, M. de la Rasilla, J. Fortea, A. Rosas, M. Stoneking, T. Schoeneberg, J. Bertranpetit, and M. Hofreiter, 2007, A melancortin 1 receptor allele suggests varying pigmentation among Neanderthals, Science 318: 1453-1455.

Lander, E. S., L. M. Linton, B. Birren,et al. of the International Human Genome Sequencing Consortium, 2001, Initial sequencing and analysis of the human genome, Nature 409: 860-920.

Lartigue C., J. I. Glass, N. Alperovich, R. Pieper, P. P. Parmar, C. A. Hutchison 3rd, H. O. Smith, and J. C. Venter, 2007, Genome transplantation in bacteria: changing one species to another, Science 317: 632-638.

Lawrence, Bruce B., 1995, Defenders of God: The Fundamentalist Revolt Against the Modern Age, University of South Carolina Press, 340 pp.

Lai, C. S., S. E. Fisher, J. A. Hurst, F. Vargha-Khadem, and A. P. Monaco, 2001, A forkhead-domain gene is mutated in a severe speech and language disorder, Nature 413: 519-523.

Lee, B. Y. and A. B. Newberg, 2005, Religion and health: a review and critical analysis Zygon 40: 443-468.

Lenggenhager, B., T. Tadi, T. Metziger, and O. Blanke, 2007, Video ergo sum; manipulating bodily self-consciousness, Science 317: 1096-1099.

Lewis, C. S., 1942, Mere Christianity, HarperSanFrancisco, reprinted 2001, 227 pp.

Malin, M. C., K. S. Edgett, L. V. Posiolova, S. M. McColley, and E. Z. Dobrea, 2006, Present-day impact cratering rate and contemporary gully activity on Mars, Science 314: 1573-1577.

Marsh, Charles, 2005, Wayward Christian Soldiers, New York Times, January, 20th.

Martin, James, 2006, The Meaning of the 21st Century, A vital blueprint for ensuring our future, Transworld Pub., London, U.K., 526 pp.

Meacham, Jon, 2006, American Gospel, Random House, New York, 399 pp.

Mikkelsen et al., 2005, Initial sequence of the chimpanzee genome and comparison with the human genome, Nature 437: 69-87.

Miller, Kenneth R., 1999, Finding Darwin's God, HarperCollins, New York, 338 pp.

Murray, Ian H., 1990, D. Martyn Lloyd-Jones: The Fight of Faith, 1939-1981, Edinburgh: Banner of Truth.

Nettelhorst, R. P. 2006, Notes on the founding fathers and the separation of church and state, www.theology.edu/journal/volume2/ushistor.htm, p. 1-8.

Nickell, J., 1998, Inquest on the Shroud of Turin: Latest Scientific Findings, Prometheus Books, Amherst, New York, 186 pp.

Nikaido, M., A. P. Rooney, and N. Okada ,1999, Phylogenetic relationships among cetartiodactyls based on insertions of short and long interspersed elements: hippopotamuses are the closet extant relatives of whales, Proceedings of the National Academy of Science, 96: 10261-6.

Noll, Mark A., 1994, The Scandal of the Evangelical Mind, Eerdmans Publ., Inter-Varsity Press, Grand Rapids, 274 pp.

Norris, P. and R. Inglehart, 2004, Sacred and Secular Religion and Politics Worldwide, Cambridge University Press, 329 pp.

Nowak, M. A., 2006, Five rules for the evolution of cooperation, Science 314: 1560-1563.

Osarchuk, M. and S. J. Tatz, 1973, Effect of induced fear of death on belief in an afterlife. Journal of Personality and Social Psychology, 27: 256-260.

Oz, Amos, 1993, In the Land of Israel, translated by Maurice Goldberg-Bartura, Harvest Books, Fort Washington, Pennsylvania, 304 pp.

Pagden, A. R., 1975, The Maya, Diego de Landa's Account of the Affairs of Yucatan (1552) J. Philip O'Hara publisher, Chicago, Illinois, 191 pp.

Pagels, Elaine, 1979, The Gnostic Gospels, Vantage Books, New York, 182 pp.

Pagels, Elaine, 1995, The Origin of Satan, Random House, New York, 214 pp.

Paley, William, 1803, Natural Theology, Oxford University Press, 2006, New York, 384 pp.

Park, Robert, 2000, Voodoo Science, Oxford University Press, 230 pp.

Patterson, N., D. J. Richter, S. Gnerre, E. S. Lander, and D. Reich, 2006, Genetic evidence for complex speciation of humans and chimpanzees, Nature 441, 1103-1108.

Paul, G. S., 2005, Cross-national correlations of quantifiable society health with popular religiosity and secularism in the prosperous democracies, Journal of Religion and Society, 7: 1-17.

Perry, G. H., N. J. Dominy, K. G. Claw, A. S. Lee, H. Fiegler, R. Redon, J. Werner, F. A. Villanea, J. L. Mountain, R. Misra, J. P. Carter, C. Lee, and A. C. Stone, 2007, Diet and the evolution of human amylase gene copy number variation, Nature Genetics 39: 1256-1260.

Phillips, Kevin, 2006, American Theocracy, Viking Press, New York 462 pp.

Pickthall, Marmaduke, 1930, The Meaning of the Glorious Koran: An Explanatory Translation, George Allen and Unwin, London.

Pinker, Steven, 2002, The Blank Slate, The Modern Denial of Human Nature, Penguin, New York, 509 pp.

Pollard, K. S., S. R. Salama, N. Lambert, M-A. Lambot, S. Coppens, J. S. Pedersen, S. Katzman, B. King, C. Onodera, A. Siepel, A. D. Kern, C, Dehay, H. Igel, M. Ares, Jr., P. Vanderhaeghen and D. Haussler, 2006, An RNA gene expressed during cortical development evolved rapidly in humans, Nature 443: 167-172.

Poser, Charles M. and George W. Bruyn, 1999, An Illustrated History of Malaria, Parthenon Publishing Group, London, 165 pp.

Price, Robert. M., 2006, The Reason Driven Life, Prometheus Books, Amherst, New York, 363 pp.

Purves, William K., G. H. Orians, and C. Heller, 1995, Life: The Science of Biology, 4th edition, Sinauer Associates, Sunderland, Massachusetts, 1195 pp.

Redon, R. and 42 additional authors, 2006, Global variation in copy number in the human genome, Nature 444: 444-454.

Rigas, Jim, 2004, Christianity Without Fairy Tales, Professional Press, Chapel Hill, North Carolina, 472 pp.

Samanta, M. P., W. Tongprasit, S. Istrailk, R. A. Cameron, Q. Tu, E. H. Davidson, and V. Stolc, 2006, The transcription of the sea urchin embryo, Science, 314: 960-962.

Schultz, Robert C., 1967, "Sermon on keeping children in school", in: Luther's Works, American ed. Vol 46, Philadelphia: Fortress, p 211.

Shu W., J. Y. Cho, Y. Jian, M. Zhang, D. Weisz, G. A. Elder, J. Schmeidler, R. De

Gasperi, M. A. Gama Sosa, D. Rabidou, A. C. Santucci, D. Perl, E. Morrisey, and J. D. Buxbaum, 2005, Altered ultrasonic volcalization in mice with a disruption in the *Foxp2* gene, Proceedings of the National Academy of Sciences, 102: 9643-9648.

Shubin, N. H., E. B. Daeschler, and F. A. Jenkins, Jr., 2006, The pectoral fin of *Taktaalik roseae* and the origin of the tetrapod limb, Nature 440: 764-771.

Sipe, Richard A. W., 1995, Sex, Priests, and Power: Anatomy of a Crisis, Brunner/Mazel, Philadelphia, 220 pp.

Smith, Huston, 2001, Why Religion Matters, HarperSan Francisco, 290 pp.

Smith, Morton, 1974, The Secret Gospel, Victor Gollancz, London, 148 pp.

Spilka, B., R. W. Hood, Jr., B. Hunsberger, and R. Gorsuch, 2003, The Psychology of Religion, an Empirical Approach,3rd edition, The Guilford Press, New York, 671 pp.

Spitzer, N. C., Electrical activity in early neuronal development, Nature 444: 707-712.

Stamatoyannopoulos, George, Phillip W. Majerus, Roger M. Perlmutter, Harold Varmus, Eds., 2001, The Molecular Basis of Blood Disorders, 3rd edition, W. B. Saunders, New York, p. 694-695.

Stark, R. and W.S. Bainbridge, 1986, The Future of Religion, Berkeley, University of California Press, Berkeley, 600 pp.

Stine, Gerald J., AIDS Update 2003, Prentice Hall, Upper Saddle River, New Jersey, 555 pp.

Swanson, Guy E., 1960, The Birth of Gods, University of Michigan Press, Ann Arbor.

Ruse, Michael, 2001, Can a Darwinian be a Christian? The Relationship Between Science and Religion, Cambridge University Press, 242 pp.

Tabor, James M., 2006, The Jesus Dynasty, Simon and Schuster, New York, 363 pp.

Tedlock, Dennis, 1996, Popul Vuh: The Definitive Edition of the Mayan Book of the Dawn of Life and the Glories of, Simon and Schuster, New York, 384 pp.

Thewissen, J.G.M., E. M. Williams, L. J. Roe, and S. T. Hussain, 2001, Skeletons of terrestrial cetaceans and the relationship of whales to artiodactyls, Nature 413: 277-281.

Thomas, Lewis, 1983, Late Night Thoughts on Listening to Mahler's Ninth Symphony, Viking Press, New York, 168 pp.

Tishkoff, S. A., and 18 others, 2007, Convergent adaptation of human lactase persistence in Africa and Europe, Nature Genetics 39: 31-40.

Trivers, R., 1971, The evolution of reciprocal altruism, Quarterly Review of Biology, 46: 35-57.

Van Doren, Carl, 1938, Benjamin Franklin, The Viking Press, New York, p. 777.

Voight, B. F., S. Kudaravalli, X. Wen, and J. K. Pritchard, 2006, A map of recent positive selection in the human genome, Plos-Biology, 4: 0446-0458.

Warraq, Ibn, Ed., 2003, Leaving Islam, Prometheus Books, Amherst, New York, 471 pp.

Warren, Rick, 2007, The Purpose-Driven Life: What on Earth Am I Here For? Zondervan Publishing, 334 pp.

Watt, W. Montgomery, translator, 1953, The Faith and Practice of Al Ghazzali, London. pp @

Weatherford, Jack, 2004, Genghis Khan and the Making of the

Modern World, Three Rivers Press, New York, 312 pp.

White, Andrew D., 1896, A history of the warfare of science with theology in Christendom, reprinted in 1993 by Prometheus Books, Buffalo, New York. 889 pp.

Wolf, Y. I. and E. V. Koonin, 2007, On the origin of the translation system and the genetic code in the RNA world by means of natural selection, exaptation, and subfunctionalization, Biology Direct 2: 14.

Wright, N. T., 2006, Simply Christian, Why Christianity Makes Sense, HarperSanFrancisco, 240 pp.

Wuthnow, Robert, 2004, Saving America? Princeton University Press, 354 pp.

Index

C

D

E

F

G

H

I

J

L

M

N

O

P

Q

R

S

T

U

V

W

Y

ISBN 142516772-1

9 781425 167721